EFFECTS OF DDT ON MAN AND OTHER MAMMALS : I.

Papers by
Thomas Jukes, Homer R. Wolfe, Donald P. Morgan, Clifford Roan, Alan Poland, Wayland J. Hayes, Jr., W.A. Gilpin, George M. Woodwell, L.J. Rogers, James L..Cox et al.

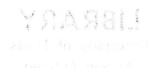

MSS Information Corporation
655 Madison Avenue, New York, N.Y. 10021

Library of Congress Cataloging in Publication Data
Main entry under title.

Effects of DDT on man and other mammals.

 Vol. 2 by Gary L. Henderson, S. M. Sieber,
W. L. Heinrichs, et al.
 1. DDT (Insecticide)--Toxicology--Addresses,
essays, lectures. I. Jukes, Thomas Hughes, 1906-
II. Henderson, Gary L.
RA1242.D35E34   632'.951   73-289
ISBN 0-8422-7110-4 (v. 1)

TABLE OF CONTENTS

CREDITS AND ACKNOWLEDGEMENTS

Cox, James L., "DDT Residues in Marine Phytoplankton: Increase from 1955 to 1969," *Science*, 1970, 170:71-73.

Gilpin, W.A., "A Case of Human Pesticide Poisoning," *Michigan Medicine*, 1970, 69:485-488.

Hayes, Wayland J., Jr.; William E. Dale; and Carl I. Pirkle, "Evidence of Safety of Long-Term, High, Oral Doses of DDT for Man," *Archives of Environmental Health*, 1971, 22:119-135.

Jukes, Thomas H., "DDT, Human Health and the Environment," *Environmental Affairs*, 1971, 1:534-564.

Jukes, Thomas H., "Fact and Fancy in Nutrition and Food Science," *Journal of the American Dietetic Association*, 1971, 59:203-211.

Jukes, Thomas H., "The Global 'Cranberry Incident'," *Clinical Toxicology*, 1970, 3:147-149.

Morgan, Donald P.; and Clifford C. Roan, "Absorption, Storage, and Metabolic Conversion of Ingested DDT and DDT Metabolites in Man," *Archives of Environmental Health*, 1971, 22:301-308.

Morgan, Donald P.; and Clifford C. Roan, "Chlorinated Hydrocarbon Pesticide Residue in Human Tissues," *Archives of Environmental Health*, 1970, 20:452-457.

Poland, Alan; Donald Smith; R. Kuntzman; M. Jacobson; and A.H. Conney, "Effect of Intensive Occupational Exposure to DDT on Phenylbutazone and Cortisol Metabolism in Human Subjects," *Clinical Pharmacology and Therapeutics*, 1970, 11:724-732.

Roan, Clifford C.; Donald P. Morgan; and Emmett H. Paschal, "Urinary Excretion of DDA Following Ingestion of DDT and DDT Metabolites in Man," *Archives of Environmental Health*, 1971, 22:309-315.

Rogers, L.J.; W.J. Owen; M.E. Delaney , "Sites of Inhibition of Photosynthetic Electron Transport by 1,1,1-trichloro-2,2-bis-(p-chlorophenyl) ethane (DDT)," *Second International Congress on Photosynthesis*, 1971, pp. 689-699.

Wolfe, Homer R.; and John F. Armstrong, "Exposure of Formulating Plant Workers to DDT," *Archives of Environmental Health*, 1971, 23:169-176.

Woodwell, George M.; Paul P. Craig; and Horton A. Johnson, "DDT in the Biosphere: Where Does It Go?," *Science*, 1971, 174:1101-1107.

PREFACE

This collection of recent papers (published 1970-1972) provides the latest information on the effects of DDT in humans and in other mammals. DDT is reported to display various toxic and physiological effects in humans. Among the reported effects of DDT are its influences on mammalian endocrine systems concerned with reproduction and its stimulation of various liver enzymes. Papers dealing with the degenerative metabolism of DDT in mammalian tissues and with aspects of DDT in the ecosystem are also included.

First in a multivolume collection on DDT, this volume is part of MSS' continuing series on environmental studies.

Occurrence of DDT in Man

DDT, HUMAN HEALTH AND
THE ENVIRONMENT

By Thomas H. Jukes

For the past two years there has been an intensive campaign to ban the insecticide DDT (dichlorodiphenyltrichloroethane) from use in, manufacture in, and export from the United States. This article evaluates the evidence for some of the charges in this campaign. These include charges that DDT is a "biocide" (a name implying that it is poisonous to all forms of life), permeates the environment, is virtually indestructible, kills many desirable forms of wildlife, and is dangerous to human beings. The last of these claims is particularly astonishing in view of the fact that DDT has saved millions of human lives from malaria, typhus fever, plague and other deadly diseases without harming a single person, except in a few cases of accidental or suicidal overdosage.

The propaganda against DDT has been so extensive and successful that DDT is now widely regarded by the public as a dangerous poison. There have been a series of newspaper cartoons, skillfully staged television shows, and a display of bumper stickers with skull and cross-bones, all aimed at exposing the evils of DDT. The National Audubon Society printed 700,000 copies of a leaflet urging that the export of DDT be stopped. This leaflet was distributed at about the same time a resolution requesting the continuation of the use of DDT was passed at a meeting of the WHO Regional Committee for Southeast Asia held in 1969 in Nepal, attended by representatives of eleven tropical countries totaling over 700 million in population—about 1000 people for each Audubon leaflet.

Many people have become self-constituted authorities on DDT as a result of exposure to mass media. For example, the chief judge of the Circuit Court of Appeals in Washington, D.C., announced in January, 1971, that DDT kills honeybees and is dangerous to people. Yet, while other insecticides, such as parathion and Sevin, do kill bees, DDT's effect on these insects is only minimal. As regards DDT's effects on people, the principal consequence has been to increase population, not endanger lives.

What are the facts of this strange paradox? What motivates those who crusade against the most useful chemical in history? Is the attack on DDT partly directed against its role in accelerating the population explosion?

The organizations that are most active in the movement to ban DDT include some of the large conservation groups. Despite the size of some of these groups, they do not speak for all segments of the population. Some environmental groups, in fact, have recently been challenged by organizations which represent racial minorities. As the following statement suggests, the needs of the urban poor are not likely to be assuaged by the Thoreau-like preoccupations of many of the conservation associations:

> In general, black people probably do not know much about the science of ecology or the study of human conservation as now offered in most universities. Probably they know very little about bay fill, polluted streams, soil erosion or redwood trees. And probably they couldn't care less. But precisely because they are black and poor, they do know a great deal about 125th Street in New York, about South Street in Philadelphia and the Fillmore in San Francisco. About those environmental disasters they are very knowledgeable and they can also tell you quite a bit about human conservation as practiced on any of those nearby street corners, in the filthy two-, three-room walk-ups, in the fetid housing projects or in their urine-, vomit-, whiskey-, blood-stained hallways which exist all over, everywhere, in the uninhabitable cities of America's enraged and inconsolable slum communities.[1]

The National Audubon Society, which appears to have a predominantly white and middle-class membership, is one of the most active anti-DDT organizations. For a member to condone the use of pesticides would be tantamount to the deepest heresy in a religious sect. For an official of the Society to approve such use would be fiscal lunacy, in view of the tremendous amount of free publicity that the Society has received as a result of *Silent Spring*[2] and other publications which have established a new mythology—the extermination of wild birds by agricultural pesticides. The Society shows underlying resentment of human beings and all their works, including cities, farms, highways, and especially private industry. Membership in the Society is a form of expiation of the sin of being one of the human race, the species that consumes "the environment". The Society stated recently that one of its two main purposes is "the education of man regarding his relationship with and his place within the natural environment as an ecological system."[3] This pious pronouncement is actually intended to exclude man as an inhabitant of the Earth, except in small numbers and in a primitive, mythical, aboriginal state. The Audubon Society has no program for the relief of suffering among millions of human beings in the tropics.

Two other organizations that have attacked DDT are the National Geographic Society[4] and the Sierra Club. The National Geographic Magazine advertises plush overseas tours for the wealthy people. The magazine has beautiful photographs of wild animals, birds, and under-dressed natives in picturesque attitudes. These pictures do not show the ravages of tropical diseases that can be controlled by DDT. The Sierra Club, which is seeking legal action to obtain a ban on DDT,[5] also features expensive outings to remote lands, again largely for the healthy and economically secure. The motivation of conservation organizations is primarily to protect the landscape and its wildlife. This may be in conflict with combatting hunger and disease in human beings.

Why is it necessary to defend DDT? Why can't other insecticides be used instead? The answer is that DDT is specifically needed to protect millions of people in tropical countries from death by malaria. This has repeatedly been made plain by the World Health Organization in statements such as the following:

> The withdrawal of DDT would mean the interruption of most malaria programs throughout the world. . . . DDT used as a residual spray of the interior surface of houses. . . led to the idea of nationwide malaria control campaigns including the whole of the rural areas of a country. The success of these campaigns resulted in the concept of malaria eradication which was adopted. . . for the world by the Eighth World Health Assembly in May, 1955.
>
> Since then DDT has been the main weapon in the world-wide malaria eradication program. Research has continued for the development of other methods of attack against malaria and for the development of alternative insecticides. To date, there is no insecticide that could effectively replace DDT which would permit the continu-

ation of the eradication program or maintain the conquests made so far.

The withdrawal of DDT will therefore represent a regression to a malaria situation similar to that in 1945. The reestablishment of malaria endemicity would be probably attained following a period of large-scale outbreaks and epidemics which would be accompanied by high morbidity and mortality due to loss of immunity by population previously protected by eradication programs.

Toxicological observation of spraymen working for a number of years in malaria eradication, and even in formulation plants, has not revealed toxic manifestations in them or in people residing in houses that have been repeatedly sprayed at six month intervals.

We therefore believe that a great harm will result from the unqualified withdrawal of DDT. We feel that selective use of DDT is justified and warranted.[6]

This is what the argument is all about. If the manufacture and export of DDT are banned in the United States, the world-wide antimalarial program will collapse. Most of the DDT manufactured in the United States is for this program. Furthermore, a ban in the United States would lead to prejudice against the use of DDT elsewhere.

Those who are fighting the ban are struggling to save lives. The objective is not to "protect the chemical industry," since the substitute insecticides are more expensive and more profitable than DDT. These substitutes can be used, with varying degrees of lower efficiency, against the agricultural pests that are controlled by DDT. But there is no effective substitute for DDT in the world-wide campaign against malaria. The other compounds either decompose rapidly, produce resistance too fast, or they are too poisonous to people.

TOXICOLOGY

There is a saying among toxicologists that this subject can be easily learned in two lessons—each five years long.

One of the oldest principles in toxicology was stated by Paracelsus almost 500 years ago: "Everything is poisonous, yet nothing is poisonous." This is quite familiar to biochemists, who recognize that several chemical elements commonly regarded as poisonous are essential in small amounts to life. Examples of these are copper, chromium, manganese and selenium. The last named of these is also carcinogenic (i.e., tends to produce cancer). Traces of practically all the elements can be detected by spectroscopic tests in most biological materials, and all living creatures contain radioactive carbon and radioactive potassium. The crucial matter is the quantity of such substances that we consume in proportion to the amount that is toxic. It is comparatively easy to poison animals with table salt in high dosage.

The development of modern analytical methods of sensitivity has enabled many substances to be detected in concentrations of less than one part per billion (ppb). If the substance thus detected is commonly regarded as a poison, then the detection of such traces in unexpected places may cause public alarm, especially when the news is presented in a sensational manner. The claimed detection of DDT in Antarctic penguins at levels in the range of one part per billion has been used to imply that the whole world has been poisoned. The finding actually shows that molecules can be dispersed widely, and that the analytical device known as vapor phase chromatography, or electron capture, is extraordinarily sensitive. Such procedures can easily give erroneous results if substances are present which simulate the compound whose measurement is sought. The analysis must be carried out by an expert, or better, by two or three experts working independently, if reliable results are to be obtained.

Another basic concept of toxicology was stated by Lucretius two thousand years ago: "Quod cibus est aliis; aliis est acre venenum"—"One man's meat is another man's poison." The scientific application of this proverb is known as comparative toxicity; it is a keystone of therapeutic medicine. A significant goal of researchers in the area is easily understood in terms of the metaphor of the "magic bullet."[7] Drug therapists have continually searched for chemical "magic bullets" which hit disease carriers while missing patients. DDT is such a "magic bullet" because it kills mosquitoes and other insects that carry disease, but does not injure human beings. The idea of the "magic bullet" is more precisely expressed by the therapeutic ratio, which is the fraction of the minimum lethal dose of a drug that is therapeutically effective. The smaller the fraction, the safer the compound is for general use.

Research in comparative toxicity has led to the discovery and development of chemicals, such as antibiotics and pesticides, which enable the human species to compete successfully with other forms of life, especially disease organisms. Although it is *possible* that advances in comparative toxicity will contribute to a serious overpopulation in man, it is *certain* that inattention to this field of knowledge will permit a critical fall in global food supplies and global public health.

Malaria

From time immemorial human beings have been hosts to a genus of protozoa known as *Plasmodia*. There are many species in this genus; three of the principal ones that attack human beings are *Plasmodium vivax*, *P. malariae*, and *P. falciparum*. They produce respectively three different types of malaria. Of these, the most malignant is that produced by the *falciparum* species, which often attacks the brain. These three Plasmodia spend part of their life cycle in mosquitoes, but the cycle is not complete without going through several stages in man, where they reach maturity in the red blood cells and reproduce, in enormous numbers, into a form called *merozoites*. These change into the sexual stage, which enters the body of a blood-sucking mosquito. Other stages of the life cycle then take place, and the parasite reaches the salivary gland of the mosquito, from which it is inoculated into the next victim bitten by this insect. The cycle then continues from man to mosquito, and mosquito to man. The principal method for breaking this pernicious chain is to kill mosquitoes with DDT by spraying the interior walls of human dwellings.

Public health authorities in the tropics apparently use a figure of 1 per cent per year to estimate the mortality from malaria; thus 75 million cases in India in one year were calculated to be responsible for 750,000 deaths.[8] The survivors in many cases are severely debilitated and unable to work.

The mosquitoes rest on the walls by day and attack sleeping people at night. If DDT is sprayed on the walls, it kills the mosquitoes. An insecticide for this purpose must be *persistent* because it is not possible for spray teams to go into the same house frequently. Malathion, for example, is non-persistent and soon decomposes; it also has a pronounced and offensive odor. The procedure must be *safe* to those dwelling in the house, and to the sprayers, who are intensively exposed to the insecticide. The development of resistance by mosquitoes to the insecticide must be slow enough to enable the life cycle of the mosquito to

15

be broken. DDT as opposed to lindane, for example, fulfills this requirement. Note that the WHO program favors spraying walls to kill adult mosquitoes rather than spraying ponds and swamps to kill larvae.

I have not mentioned cost, because this should not be a prime consideration when many human lives are at stake. However, Russell estimates that the appropriations of the United States for overseas malaria control and eradication amounted to a half-billion dollars in the past 25 years.[9] As a result of international cooperation, WHO has maintained an effective world-wide malaria eradication program:

> Today more than 960 million people who a few years ago were subject to malaria endemicity are now free of malaria; another 288 million live in areas where the disease is being vigorously attacked and transmission is coming to an end. Because much of Africa remains highly malarious and because about 288 million people live in malarious areas not yet subject to eradication measures, it is logical that the United States should maintain an active interest in this disease.[10]

These estimates by Dr. Russell indicate that the United States contribution to saving lives from malaria has paid off quite well in terms of human welfare. As I said recently however, "some Americans, by demanding a ban on DDT, are reversing the traditional role of their country in relieving the sufferings of others."[11]

DDT in Agriculture

Insects compete with mankind for food. They devour all parts of plants—leaves, stems, fruits and seeds. A plant attacked by insects will often die without producing seeds or fruit. The vulnerability of plants to attack by insects is greatly increased by agriculture, which inevitably leads to what is called "monoculture"—the growing of a single crop in a large area of land, such as a field. Obviously a group of crop plants, such as potatoes, corn, tomatoes and alfalfa, cannot be grown as a mixture. Food must be grown on farms using monoculture, unless and until some other method of providing nourishment is developed.

Prior to World War II methods of controlling agricultural pest insects included the use of "stomach poisons", the most effective of which was lead arsenate; the use of contact poisons, e.g., nicotine sulfate; and the use of hand labor, e.g., for burning cornstalks and plowing them underground in the fall to control the European corn borer, or for picking bugs off potato plants. The

16

discovery of DDT revolutionized the control of agricultural insects. It also replaced lead arsenate, which is virtually indestructible, and highly toxic.

DDT is intensely poisonous for many insects, and is less toxic for plants and warm-blooded animals. Like all chemicals, however, DDT has a level of toxicity for any species of animal or plant. The extensive use of DDT, without adequate controls, has resulted in the killing of non-target species. In some instances this was an accepted risk. For example, when forests were sprayed to control the spruce budworm, many fish were killed in the streams. These were replaced by planting other fish.

DDT is poisonous to crustacea, such as crabs. Careless use of DDT and other pesticides that results in their drainage into rivers, swamps, lakes, estuaries and coastal waters must not be allowed. This precaution would remove a major source of friction between agriculture and those who are interested in wildlife. Actually, a rapid reduction is currently taking place in the agricultural use of DDT, and this will undoubtedly lead to a lessening of its movement through "food chains," which concentrate fat-soluble substances. However, if DDT suddenly disappeared this would not end the problem. Most of the fish that are killed by pollutants in inland waters are the victims of petroleum wastes, industrial wastes and sewage, just as most eagles found dead are killed by gunfire. The role of other pollutants is just beginning to emerge.

Solubility in fats is not an exclusive property of DDT or of the other pesticides that are similar to it. Recently it has been found that mercury can be converted into a fat-soluble form, methyl mercury, by bacteria. This form can enter food chains and, by stepwise concentration, produce effects that are toxic to animals and birds that eat fish. Mercury has been present in sea water ever since the oceans were formed, and occurs in the earth's crust. It does not become a problem until its concentration is greater than the toxic level. This level can result from industrial contamination, but the possibility exists that such levels also occur under natural conditions.

Space does not permit an adequate discussion of the vast topic of the effect of chemicals on wildlife. Many of the conclusions are based on inferences, rather than on controlled experiments. There arise therefore ambiguities, which lead in turn to disputes. The final answer may differ from the first guess. It is thus best to react cautiously to preliminary judgments regarding the death or disappearance of wildlife. For example, a few winters ago robins were unusually scarce in the coastal cities of California and DDT was blamed widely. The robins were actually back in the mountains, feeding on an unusually fine crop of berries, and the following winter they were in town as usual.

How to Alarm the Public—A Study in "Eco-Tactics"

A syndicated newspaper article has described the history and activities of the Environmental Defense Fund (EDF), asserting that the EDF "has swiftly become the public defender of the environmental movement." The article stated that

> The turning point came when Cameron decided to spend about $5,000 of the organization's total remaining assets of $23,000 on an advertisement in the New York Times on Sunday, March 29, headlined 'Is Mother's Milk Fit for Human Consumption?' It referred to the amounts of DDT in the human body.
>
> The ad appealed for members, starting at $10 for a basic membership. It produced $7,000, a profit, and the EDF turned to a direct mail campaign and now has 10,000 members, a stable financial base and a chance at major foundation support.[12]

This is a most interesting revelation. The EDF appealed to the public on the basis of the DDT content of human milk. As a means of arousing alarm concerning DDT, the EDF and the National Audubon Society have both stated that DDT causes cancer. The implication that DDT in breast milk may cause cancer in babies is superlatively sensational copy. The following lurid passage is from an article by Ed Chaney, Information Director, National Wildlife Federation:

> A five-day-old human being lies asleep in the other room. His name is Eric. His tiny, wiggly, red body contains DDT passed on to him from his mother's placenta. And every time he sucks the swollen breasts, he gets more DDT than is allowed in cow's milk at the supermarket. Be objective? Forget it. Objective is for fence posts. How can you be objective in the face of a global insanity that is DDT? In the face of abdicated responsibility by the men the public pays to protect its interests. Are the anarchists right? Are ashes the only fertile seed bed for growing new responsiveness to the public interest? Picture a swarm of angry citizens bathed in the light of flames engulfing the Agriculture Department.[13]

It is distressing that an official of a large organization should discard objectivity and propose anarchy in its stead.

Let us examine the factual and scientific background for the propaganda campaign regarding DDT in human milk. The background starts with the improvements in technology that made it possible to detect fantastically small quantities of DDT. Note that such extremely delicate tests can easily give "false positive" readings because of accidental contamination of the equipment or lack of expertise by the tester. We must next note that cows' milk has occupied an unusual position among foods with respect to regulations. "Zero tolerance" has been the policy with respect to additives to milk, except for vitamin D. The improvements in testing procedures made it necessary to re-examine the definition of zero, since every chemist knows that zero content, in molecular terms, does not exist. For example, all fish and all human beings have contained mercury for millions of years (*i.e.*, before the chemical industry existed). To get back to milk, more than ten years ago it was evident that the entire canned milk stocks of the United States contained DDT. It was therefore necessary to face facts and choose one of two alternatives: ban cows' milk from interstate commerce, or set a tolerance.

The second alternative was chosen, and the tolerance set at 0.05 parts per million (ppm). This was a far lower level than the 7 ppm permitted for most agricultural products. A rule of thumb for tolerance levels is 1% of the toxic dose which is lethal to 50% of a group of experimental animals. Obviously, if 7 ppm had been estimated to be non-injurious, a tolerance of 0.05 ppm provided an unusually large margin of safety. The low tolerance was possible primarily because cows metabolize and break down DDT very effectively, and also because great attention was paid to avoiding the use of DDT on crops, such as alfalfa, which are consumed by dairy cattle. In contrast, human beings are less efficient than cows in metabolizing DDT, and they do not eat hay. There is a straight-line relationship between DDT intake and DDT level in body fat.[14] If the dosage decreases, the content of DDT in the fat becomes less. This is the result of an equilibrium level between intake, breakdown and excretion in the urine. A level of 10 ppm in the body fat is apparently harmless; far higher levels occur in spray operators and workers in DDT manufacturing and formulating factories who remain in good health despite prolonged exposure for periods up to 20 years.[15] Dr. J. M. Barnes (Director, Toxicology Research Unit, British Medical Research Council) summed up the matter as follows:

> Unfortunately, DDT is relatively slowly metabolised and excreted by the mammal and by virtue of its solubility characteristics tends to get laid down in tissue fat. Here it would have remained as an innocent and unrecognised passenger but for the fact that the chemists invented a sensitive chemical method, since further enhanced by the gas-liquid-chromatographic technique capable of detecting the chlorine and indicating its source even in minute quantities. Thus it has become possible to establish an anxiety neurosis in respect to a few parts per million of a compound in a tissue such as fat where a few parts per thousand in the whole animal are of no toxicological significance.[16]

A good example of one of the many studies on the prolonged effects of DDT on human subjects is the recent publication by Hayes and co-workers who reported that:

> Twenty-four volunteers ingested technical or p,p'-DDT at rates up to 35 mg. per man per day for 21.5 months. They were then observed for an additional 25.5 months, and 16 were followed up for five years. Storage of DDT and DDE and excretion of DDA were proportional to dosage. The fat of those receiving technical insecticide at the highest rate contained 105 to 619 ppm of DDT when feeding stopped. The average dosage of p,p'-DDT administered in this study was 555 times the average intake of all DDT-related compounds by 19-year-old men in the general population and 1,250 times their intake of p,p'-DDT. Since no definite clinical or laboratory evidence of injury by DDT was found in this study, these factors indicate a high degree of safety of DDT for the general population.[17]

DDE and DDA are two breakdown products of DDT. DDE is not insecticidal, but it has an effect similar to that of DDT in inducing the production of microsomal enzymes. DDA is an acetic acid derivative. It is inert, is soluble in water and is excreted in the urine.

The DDT in human beings enters the fat of breast milk. This was noted and published in 1950 by Laug and co-workers who found an average concentration of DDT of 0.13 ppm in 32 samples taken in Washington, D.C., with a range from undetectable ("zero") to 0.77 ppm.[18] Several similar reports have since appeared, and the results of an extensive survey were described by Quinby and co-workers.[19]

It may be concluded from the preceding discussion that the DDT level in human milk is about twice as high as the tolerance allowed for cows' milk by the FDA. That bald conclusion, however, requires explanation: it must be explained in terms of its underlying premises and toxicological implications. The use of the unqualified conclusion to create public alarm is a scientifically irresponsible act.

The DDT content of human milk has also been scrutinized by the World Health Organization and the Food and Agricultural Organization of the U.N. They set a permissible rate of intake of 0.01 mg. of DDT per kilo of body weight for breast-fed infants. The DDT intake of breast-fed babies in the United States may be higher than this; estimates range from 0.014 and 0.02 mg/kilo/day at birth, if the infant consumes 600 ml. (about 1⅓ pints) of breast milk daily. As the infant grows the intake of milk on a per-kilo basis decreases because food intake per unit of body weight lessens when the size of an animal increases. Furthermore, breast-fed infants usually receive supplementary feeding with other foods.

The "permissible rate" set by the WHO-FAO, according to the chairman of the meeting that established the value, is highly conservative, and he points out that

it offers a safety factor of about 25 compared with what workers in a DDT manufacturing plant have tolerated for 19 years without any detectable clinical effect (see Laws *et al.*, *Arch. Environ. Health*, 15: 766–775, 1967). The safety factor of the WHO-FAO permissible rate is 150 compared to the dosage of DDT given daily for 6 months to a patient with congenital unconjugated jaundice without producing any side effects (Thompson *et al.*, *Lancet* 11, (7610): 2–6, July 5, 1969).

Infants are more susceptible than adults to some compounds, but the difference is seldom great—usually about 2 to 3 times. In a study of 49 different compounds, newborn rats were found to vary from 5 times less susceptible to 10 times more susceptible than adults. Although there is no information on the relative susceptibility of human infants and adults to DDT, it is shown by Lu *et al.* (*Food and Cosmetic Toxic.*, 3: 591–596, 1965) that weanling rats are slightly more resistant than adult rats to this compound, and that preweanling rats are more than twice as resistant and newborn rats are over 20 times more resistant than adults.[20]

Evidently it is possible for breast-fed infants to obtain DDT from the milk at a level up to twice the WHO-FAO "permissible rate." Again, background information indicates that no toxic effects have been detected or could be anticipated at this level. Nevertheless, the EDF and its collaborators have conspicuously proclaimed a warning that DDT may cause cancer. This adds to public apprehension, especially among nursing mothers. The question of carcinogenicity therefore should be discussed.

The "Delaney Clause" of the Food Additives Amendment[21] prohibits the use of any food additive that has been found to cause cancer in experimental animals. It is difficult to think of a more meritorious or public-spirited objective than is implied by this clause. It is even more difficult to *comply* with it, because all foods contain substances which can be shown to cause cancer in experimental animals, given the appropriate dose and the susceptible animal. All foods contain traces of radioactive elements which are present naturally. All meat products contain sterols and steroid hormones, which produce breast cancer in mice.

Pyrolysis—scorching—such as occurs in barbecuing of meat or the roasting of coffee, produces carcinogens and is only one of many examples of processes or substances in foods that can produce cancer in experimental animals when at high levels. The Delaney clause is usually regarded among scientists as being impossible either to administer or repeal. The Secretary's Commission on Pesticides, Department of Health, Education and Welfare made the following recommendation and comments:

Recommendation 8: Seek modification of the Delaney clause to permit the Secretary of the Department of Health, Education, and Welfare to determine when evidence of carcinogenesis justifies restrictive action concerning food containing analytically detectable traces of chemicals.

The effect of the Delaney clause is to require the removal from interstate commerce of any food which contains analytically detectable amounts of a food additive shown to be capable of inducing cancer in experimental animals. This requirement would be excessively conservative if applied to foods containing unavoidable trace amounts of pesticides shown to be capable of inducing cancer in experimental animals when given in very high doses. If this clause were to be enforced for pesticide residues, it would outlaw most food of animal origin including all meat, all dairy products (milk, butter, ice cream, cheese, etc.), eggs, fowl, and fish. These foods presently contain and will continue to contain for years, traces of DDT despite any restrictions imposed on pesticides. Removal of these foods would present a far worse hazard to health than uncertain carcinogenic risk of these trace amounts.

22

Commonly consumed foodstuffs, contain detectable amounts of unavoidable naturally occurring constituents which under certain experimental conditions are capable of inducing cancer in experimental animals. Yet, at the usual low level of intake of these constituents, they are regarded as presenting an acceptable risk to human health.

Exquisitely sensitive modern analytical techniques which became available since enactment of the Delaney clause permit detection of extremely small traces of chemicals at levels which may be biologically insignificant. Positive response in carcinogenic testing has often been shown to be dose-related, in that the carcinogenic response increases with increasing dose levels of the carcinogen; when the dosage of a carcinogen is minimized, the risk for cancer is also minimized or eliminated. . . .

The recommendation for revision of the Delaney clause is made in order to permit determinations essential to the protection of human health, not to justify irresponsible increases in the exposure of the population to carcinogenic hazards.[22]

However, any attempt at such a revision would meet with great political opposition in the light of current fears and superstitions regarding "chemicals."

The above quotation speaks of the "uncertain carcinogenic risk" of trace amounts of pesticides. I shall review this statement with regard to DDT. The question of carcinogenicity of DDT was examined extensively starting in 1947. The above statement by the Commission was based on a recent report by Innes et al.[23] which is essentially a repetition of observations made about 20 years ago. The extensive earlier work on DDT and tumors in experimental animals includes about 20 articles in the scientific literature. In 1944, Lillie and Smith described hepatic alterations in rats kept on a diet containing 1000 ppm of DDT for 14 weeks.[24] Similar changes were observed repeatedly by subsequent investigators.

These findings aroused much interest and a number of toxicologists studied the effects of DDT in various experimental animals. Cameron and Burgess fed very high levels of DDT to rats, and produced liver damage that was severe enough to account for death in a number of the animals.[25] However, when the DDT was discontinued "the dead cells were removed by autolysis and phagocytic action and repair was complete without any fibrosis, although calcification was occasionally seen."[26] In plain language, if the rats were fed enough DDT to produce acute liver damage, and then the DDT was stopped, the animals got better. No cancer was found.

A more prolonged ordeal for rats, which were given a level of 100 ppm of DDT in the diet, was described by Fitzhugh and Nelson.[27] After two years, which is roughly equivalent to 70 years for a human being, a "minimal hepatocarcinogenic tendency" was noted by the authors. The authors could not decide whether the tumors were benign (adenomas) or low grade hepatic cell carcinomas.

These findings draw attention to an interesting matter known as the *dose-response curve*. It is possible to calculate the time of onset of symptoms from the daily dose of a toxic substance. If the dose is low enough, a calculation may show that the average animal will die of old age before it develops tumors. Since biological responses are subject to individual variability, the "average" animal does not represent all the animals in a group. There will be a few animals, perhaps only one in a million, that will develop tumors at a considerably lower dose than the average. This is the basis for suggesting so-called "mega-mouse" experiments (a million mice per experiment!) to detect borderline effects of mutagens and carcinogens. But in a million animals there are enough spontaneous tumors and mutations to make the results undecipherable at low levels of the chemical.

Fitzhugh and Nelson also described recovery experiments with rats on 1000 ppm of DDT for 12 weeks.[28] Extensive necrotic changes in the liver were produced during this period. The livers returned to normal after 8–10 weeks, indicating an absence of malignancy in the 12-week lesions. In these and various other studies with DDT, the minimum concentration in the diet was 100 ppm. At lower levels, the smallest detectable morphologic effect was at 5 ppm, reported by Laug *et al.*[29] These authors also reported hepatic cell changes in rabbits, mice and guinea pigs fed DDT, but the changes were not as marked as in rats. They were absent from chickens, dogs, cats, monkeys and large domestic animals. As a result, the liver changes are regarded by many pathologists as being characteristic of rodents. Various investigators have been unable to produce any pathology in rats fed DDT; indeed, the British investigators Cameron and Cheng concluded that the histological changes reported by others were "fixation artefacts."[30] These differences of opinion stimulated further research. This extensive series of publications and investigations was reviewed in 1965 by Arnold Lehman, chief of toxicology for the United States Food and Drug Administration, and his conclusion was that "DDT is not a carcinogen."[31]

The liver changes produced by DDT in rats and other rodents were described by Hayes as involving the cellular tissue that produces the "microsomal enzymes", as being reversible, and as being peculiar to rodents.[32] He also pointed out that the changes could also be produced by phenobarbital, by pyrethrum (a "natural" insecticide), by ethyl alcohol and by oxidized fats. The microsomal enzymes have various effects including the breakdown of some toxic substances and certain hormones.

The allegations by the EDF that DDT is a carcinogen are based primarily on three reports. The first is an article by Innes and co-workers with mice, which the authors call a "preliminary note."[33] Just why it was necessary to publish a "preliminary note" on a subject (the effect of DDT on rodents) that had been covered exhaustively over a period of ten years by a large number of experiments is not clear. The amount of DDT fed was the maximum tolerated dose, 140 ppm in the diet, and the experiment lasted 18 months, which is most of the lifetime of a mouse. The results were as follows:[34]

Supplement (and no.)	Level used		% Mice with tumors		
	mg/kg*	ppm	Hepatomas	Lung	Lymphoma
None (controls) (338)	—	—	4.1	6.2	4.1
p,p'-DDT (72)	46.4	140	31.9	5.5	10.9

* Dosage was oral for 7–28 days, then added to diet.

In confirmation of experiments reported earlier by other scientists, there was an eight-fold increase in hepatomas over the controls, and a borderline effect on lymphoma, to which mice are highly susceptible. Hepatomas are defined as tumors that are on the borderline between benign and malignant. Dr. Hayes commented as follows:

Innes, et al. (1969) reported that the tumorigenicity of selected pesticides and industrial compounds was tested by continuous oral administration to both sexes of two hybrid strains of mice, starting at the age of 7 days. The chemicals were given by stomach tube until weaning and thereafter as a mixture in the diet. Maximal tolerated doses were given for the entire period of observation, about 18 months. The authors stressed that the dose received by the mice was far in excess of that likely to be consumed by humans. One of the compounds that gave a statistically significant positive result was DDT. The incidence of tumors was comparable to the mean tumor incidence produced by a group of positive control compounds, most of which are weak or even questionable carcinogens of no demonstrated importance to human health. The authors made no distinction between hepatomas and carcinomas. It is difficult to understand why, in denying the practicality of making this important distinction, they entirely neglected the matter of reversibility. A full account of the study is promised later. In the meantime there is no assurance that the small number of tumors observed in mice exposed to DDT were different from the "nodules" described by Fitzhugh and Nelson in 1947.[35]

In this experiment, the mice received about 3000 times as much DDT in their diet as is consumed by people in the United States. It is primarily on this experiment that the allegations rest that DDT causes cancer. The other two reports are trivial; in one of them the diet of mice was suspected of being contaminated[36] and the second was based on a far-fetched inference which one of the authors (Deichmann) later "soft-pedaled."[37] The inference was that DDT was found in higher concentrations in fatty tissues of persons with terminal degenerative diseases than in fatty tissues from "normal" autopsy samples. However, such terminal diseases are usually accompanied by emaciation, which would concentrate the DDT in the remaining fat, and Deichmann, *et al.* have commented that the "investigators did not demonstrate a causal relationship between those diseases and pesticide retention in body tissues."[38] Yet a causal relationship is repeatedly emphasized by the lawyers for the EDF, the National Audubon Society and the Sierra Club.

The argument over DDT and cancer is important because millions of lives hang on it. If DDT were officially tabbed as a carcinogen in the United States, its use in the world-wide malaria control program would be severely inhibited or even stopped. What would this do? Ceylon has provided an excellent object-lesson:

Following a country-wide malaria eradication campaign in the 1950's and early 1960's, the number of confirmed malaria cases reached lows of 31 in 1962 and 17 in 1963, when full-scale house spraying was partially withdrawn, and subsequently terminated in 1964. The cases increased annually thereafter, numbering 150 in 1964, 308 in 1965, 499 in 1966, and 3466 in 1967, most of them occurring in the last few months of that year. In 1968, the epidemic flared rapidly—16,493 confirmed cases being reported in January and 42,161 in February. No DDT supplies were on hand with which to reinstate the house spraying program on the wide scale needed, and months were required for the procurement and delivery of them. As a result, more than a million cases of malaria occurred throughout the country in 1968.[39]

What is the *real* effect of DDT on babies? Let us reverse the coin—What is the effect of malaria? Burnet had something to say on this. He wrote in 1953 that malaria

is the great devitalizer of the tropics. . . and it is the main agent of infantile mortality. If malaria could be suddenly eliminated from the globe, the racial, economic and political consequences within a very few years would probably be appalling. India and parts of Africa are populated up to and beyond the capacity of the land to provide adequate food by present methods, and even with the tremendous infantile and prenatal mortality caused by malaria, the populations are increasing steadily. The sudden conversion to a more vigorous and rapidly increasing population would undoubtedly produce famine (emigration) and intense internal and external social repercussions.[40]

Nine years after Burnet's article, the effects of the anti-malaria program in India were described by Pal:

The control programme [with DDT] was launched in April, 1953 and it was designed to give protection to 200 million people living in the malarious areas of the country. . . . Improved knowledge on malaria control led to the revision of the original strategy and the aim became the eradication of the disease for the entire sub-continent. In April, 1958, the National Malaria Eradication Programme was launched. It consists of three phases: attack, consolidation and maintenance. The attack phase is aimed at total interruption of transmission by spraying with residual insecticides all roofed structures throughout the country. . . . Since 1953, about 147,593,270 lb. of DDT have been used, with small amounts of BHC and dieldrin. As a result, malaria morbidity has been significantly reduced in the country. The proportional case rate of malaria (per cent of malaria cases to total diseases as clinically diagnosed) in each year of this programme has shown a decline. . . . Estimates of actual morbidity and mortality are difficult but it would appear, from the available data, that malaria in India has been reduced from 75 million cases to less than 5 million. A new era in economic development and social progress has been initiated with its beneficial transformation of the life of the people. The average span of life in India is now 47 years, whereas before the eradication campaign it was 32 years. This improvement has resulted in better agriculture and industrial production. In the Terai region (Uttar Pradesh), land under cultivation and food grain production has increased and this region, once abandoned by its inhabitants because of the high incidence of malaria, has become a beautiful and prosperous area.[41]

WILDLIFE, NATURE STUDY AND INFERENCES

It is often difficult to obtain controlled and reliable results on the question of toxicity of chemicals to wildlife under field conditions. The temptation to blame pesticides indiscriminately for the death of wild creatures seems irresistible to organizations committed to the protection of wild animals. A trace of DDT reported in a dead bird or fish often triggers a chain reaction of publicity and incrimination. On the other side of the ideological fence, the farmers and entomologists give considerable weight to the fact that, in order to be able to eat and to protect themselves against major diseases, human beings must vigorously wage war on noxious insects. Chemical pesticides are essential in this fight. The release of pesticides into the environment, and the presence of traces of pesticides in our food, are inevitable if the human race is to maintain its present numbers and control of disease. It is obvious, or course, that the use of pesticides must be kept down to the minimum level commensurate with adequate crop production and disease control.

Modern agricultural technology as practiced in the United States is a development that has taken place largely in the past half-century, with major advances in the last 25 years. Geneticists, chemists and engineers have made great contributions to increasing the food supply and simultaneously lifting the burden of toil from farming. The easy availability of food has accelerated the movement of people to the cities.

Food is never pure. In the 1930's, contamination of food by pests was a major problem; for example, canned vegetables were spot-checked for pieces of insects. Today much testing is done for pesticide residues. With rare exceptions, the amounts found are well below the tolerances, which in themselves are far below the toxic levels. Pesticide residues in foods are not a public health problem. The absence of such residues could be brought about by stopping the use of pesticides. This would create a real problem—food shortages.

It would not be possible to reverse this movement and replace chemical technology by hand labor on farms without a great social dislocation. If chemical herbicides are not to be used, "the

28

man with the hoe" must return to the farm and work long hours. In China, until recently, grasshoppers were killed by hand, but most of the major pest insects are too elusive for even this archaic procedure, and "biological control" can play only a minor part in keeping such insects from destroying crops.

Many of the charges that DDT destroys wildlife are based on inference. Often the charges have been based on tests that appeared to detect traces of DDT or other pesticides, and no consideration was given to the quantitative aspects of the results. Sometimes the effect of non-pesticide factors is disregarded or ignored. An example is the occurrence of large numbers of dead fish in the lower Mississippi River. This was blamed, with great fanfare, on pesticide contamination of the water. Subsequently the deaths were attributed to bacteria, *Aeromonas liquefaciens*, and to a lack of oxygen resulting from run-off of flooded fields.

The decline of the crab catch in the vicinity of San Francisco in 1969 was blamed on DDT. A scare article and banner headline appeared on the front page of the *San Francisco Chronicle* stating that the decline resulted from the toxic effects of DDT.[42] The story warned the public that crab meat might be contaminated. Strangely enough, the crab fishery further north on the California coast reported record catches for three seasons, and the most recent one was 14 million pounds (making one wonder whether the species will be "fished out"). In November, 1970, another San Francisco newspaper headlined: "What Happened to Our Crabs? Pollution!"[43] The article stated:

> The dumping of millions of gallons of highly poisonous wastes off the Farallones is probably responsible for the drastic slump in the San Francisco crab fishery.
>
> State Fish and Game Biologists Don Lollock and John Ladd in a report to the Regional Water Quality Control Board said that although absolute proof is lacking, evidence of the decline points to the oceanic pollution.
>
> The catch has dropped from a high of nearly nine million pounds in the 1956–57 season to 1.4 million last season.
>
> The biologists recommended the board take prompt action to stop such industrial waste discharging.
>
> On receiving the State Fish and Game Department document, the board yesterday took emergency procedures to place the subject on the agenda for action at its Dec. 22 San Francisco meeting.
>
> Three firms use the ocean for major dumping. U. S. Steel, given a Dec. 15 deadline yesterday for improving the quality of discharges

from its Pittsburg works, dumps in the neighborhood of 15 million gallons of acid steel waste containing sulphuric and hydrochloric acid annually.

Sulfates and large concentrations of heavy metals also are barged and unloaded in the Farallones Gulf, nine miles from the City.[44]

What happened to our crabs? What happened to the DDT that was blamed in 1969?

DDT was found in the livers of dead sea lions on the coast of California in the late summer of 1970, and, as usual, the newspapers swung into action to condemn the insecticide. In November, a memorandum was issued by Dr. Richard Hubbard of the Marine Mammal Study Center, Fremont, California, describing the findings of a team that had diagnosed the deaths as being due to leptospirosis. The diagnosis was based on symptomatology, post-mortem findings, identification of *Leptospira pomona*, blood antigen tests and epidemiology. He commented that "there is no correlation between mercury and DDT levels, and sick animals." But, by November, who was interested? Certainly not the EDF. *Leptospira* is "part of the environment."

Robinson Jeffers has written, "Give your heart to the hawk." The bird protection societies would seemingly have us follow this admonition and, indeed, the speed and audacity of the bird of prey appeal to the Walter Mitty who lurks in most of us. Eagles have many admirers, while chickens (at least, while alive) have but few. The dire warnings that bald eagles were threatened by DDT were enough to arouse horror against this insidious chemical. For example, a photograph on the cover of *Science*[45] showed a bald eagle's nest with one unhatched egg and one apparently healthy eaglet. The failure of one of the eggs to hatch was attributed to the presence of DDT in fish in the Great Lakes area. But is the story true?

An examination of scientific literature which antedates the extensive use of DDT is instructive. It reveals that even in such earlier years the survival of the eagle was deemed a critical issue. For example, in 1921 an article entitled "Threatened Extinction of the Bald Eagle" appeared in *Ecology*. In 1943, F. Thone stated in *Science News Letter*: "When the timber was cleared, it was inevitable that the eagles had to go. Moreover, the cities grew and befouled the rivers with sewage and industrial wastes. The once teeming fish population vanished."

The Territory of Alaska paid a bounty of 25 cents per claw for 115,000 bald eagles assassinated between 1917 and 1952. There are now an estimated 7000 bald eagles in Alaska, about 6% of the number that was slaughtered. Gunfire continues to be the main cause of death of bald eagles, according to United States Department of Interior figures: of 76 dead specimens examined between 1960 and 1965, 44 were shot, 7 died of impact injuries, 3 of other violent forms of death, 4 of disease or old age, and the remaining 18 of undetermined causes. More recently, the same Department has reported "poisoning from dieldrin in growing numbers of bald eagles found dead in the United States"[46] and also that mercury poisoning has been detected in bald eagles.[47] The last finding draws attention to the presence of potentially toxic levels of mercury in fish in the Great Lakes, and suggests that mercury may be the cause of reproductive failure in the bald eagles in this region.

Why was DDT blamed? As George Mallory said of a certain mountain, "Because it was there."

How are the eagles doing? According to the National Park Service, a record number of 373 bald eagles, 120 of which were immature, were counted on November 20, 1969, below Lake Macdonald in Glacier National Park and 268 (with the birds still flying in) on November 25, 1970. The percent of immatures has held "a steady 31" during the past five years. This meant, according to the Park Service, that the birds came from "areas not yet seriously affected by pesticides," an explanation which is on a logical par with stating that the birds are alive because they are not dead. Evidently pesticide poisoning plays no more than a minor role in the demise of bald eagles. The alleged effects of DDT are now being second-guessed in favor of other chemicals.

In Scotland, the use of dieldrin in sheep dips was concluded to be the cause of a decline in the breeding ability of the golden eagle, which eats sheep carrion. This use of dieldrin was discontinued in 1966, following which there was an improvement in breeding success in golden eagles.[48]

The story of the peregrine falcon is similar in many respects to that of the eagle, except that the peregrine is even rarer. It is not only the object of man's hostility because it eats racing pigeons, but it is also harassed by the robbing of its eyries to provide young birds for falconry, and eggs for collectors. The peregrine falcon is considered as no longer breeding in the eastern United States. However, the counts of migrant peregrines at Hawk Mountain, Pennsylvania, were as follows:

31

1935–1942	ave.	32
1967		22
1968		21
1969		26
1970		27

It is obvious that the peregrine was rare even prior to 1940; the total annual count of hawks at Hawk Mountain ranges between 10,000 and 20,000.

The rapid decline of the peregrine in the eastern United States took place prior to the introduction of organic pesticides,[49] and seems to have been caused by harassment.[50] The peregrine population in the eastern United States is estimated as having been something less than 275 breeding pairs in 1940. In contrast, the peregrine population in Northern Canada and Alaska is reported to be thriving and was estimated by Fyfe (1969) as about 7,500 breeding pairs.[51]

Whether or not counts like those above can be read as linking DDT with animal deaths, the counts must be kept in proper perspective. All such statistics should be reviewed in the context of other "human" statistics. One cannot ignore, for example, that before the advent of DDT, 2000 *people* per *day* were dying in India from malaria.

Other analyses have been made of the effects of DDT on the peregrine. High levels of DDT and its metabolites were reported in the fat of peregrine falcons in the Yukon[52] but a "seemingly normal average" of viable eggs and young was found in 15 nests in this region. Fat biopsy samples from nine females had an average content of 37 ppm DDT, 284 ppm DDE and 40 ppm TDE (DDD). The samples also contained an average of 3.3 ppm dieldrin and 4.4 ppm heptachlor epoxide. During this survey, the nests were robbed to obtain eggs for analysis, and female falcons were trapped and slit open to take samples of body fat. Following this, the unfortunate birds were sutured and released, after which they "showed normal behavior." One wonders how long they survived. The authors noted that the eggs were taken from eyries where reproduction was normal and averaged about 27 ppm of total organochlorine residues. Ratcliffe stated that the residues in eggs from unsuccessful eyries were 17.4 ppm; these were to be compared with the residues of 12.7 ppm in eggs from successful eyries in Great Britain.[53] The disparity between the two sets of findings suggests that the reason for reproductive failure in the peregrine has not been identified.

The British Advisory Committee on Pesticides and other Toxic Chemicals, in a comprehensive review on organochlorine pesticides published in 1969, stated:

> There is no close correlation between the declines in populations in predatory birds, particularly in the peregrine falcon and the sparrow-hawk, . . . and the use of DDT. Therefore DDT does not appear to have been the principal cause.[54]

In summary, the peregrine as a breeding species in the eastern United States appears to have been extirpated primarily by harassment, and no role for DDT has been shown. The peregrines in northern Canada are breeding successfully despite high tissue levels of organochlorine pesticides. The current reduction in the use of these pesticides should tend to a lowering of these levels, and the future of peregrines in this region does not appear to be in jeopardy.

The osprey, or fish hawk, was a conspicuous species along the eastern seaboard of the United States, even in populated areas such as Gibson Island on Chesapeake Bay. A decline in the population of ospreys in such areas has been attributed to DDT. Harassment seems more likely to be the major cause, because the osprey is evidently increasing in the eastern United States and Canada. An obvious explanation is that the birds have moved away from the region of suburban developers and outboard motors in search of peace and quiet. The osprey count at Hawk Mountain averaged 172 per year from 1935 to 1942, pre-DDT years. Recent counts are:

1965	444
1966	405
1967	467
1968	403
1969	530
1970	600 (a record high)

The counts at Hawk Mountain must be interpreted with care, because up until about 10 years ago there was only one look-out, and now there are three, at least two of which "are manned every day, all day, from mid-August through November," according to Dr. J. W. Taylor, who estimates that 20 years ago, under today's conditions, the "number of hawks seen would have been at least three times, and probably five or six times, what our figures in 1969 show." Dr. Taylor nevertheless states that "the

Osprey population on the interior lakes has greatly increased, and these are the birds we are now seeing at Hawk Mountain."

Naturally all of us hope that hawks and eagles will survive the onslaughts of human interference. However, blaming DDT seems a convenient excuse. If a species is still being counted at Hawk Mountain, it is obviously *not* extinct, and if the numbers have not decreased greatly in recent years an interpretation that the species is drastically declining is questionable in the absence of further study.

DDT AND PHOTOSYNTHESIS IN THE OCEAN

DDT is said to be steadily accumulating in the seas by distillation from the surface of the land, by drainage into rivers and by the blowing of dust. It is also alleged that DDT is, in effect, indestructible, because its principal metabolite, DDE, resembles DDT in inducing the production of microsomal liver enzymes. One of the most sensational anecdotes about DDT is the prediction that it will stop photosynthesis in the oceans, as a result of which life on Earth will become extinct. This story was repeated by diverse authorities ranging from U Thant to Ehrlich, who stated[55] that it originated in Wurster's report.[56] This report described the effects of adding graded amounts of an alcoholic solution of DDT to algal cultures in sea water and measuring photosynthesis by carbon-14 uptake. The DDT was added to produce concentrations up to 500 ppb. Not surprisingly, photosynthetic uptake of carbon-14 was depressed at the higher levels, since DDT is phytotoxic above certain levels. Its maximum solubility in sea water is 1.2 ppb, and DDT would be precipitated and adsorb to the algae above this level. The results as presented by Wurster show that at the points corresponding to 1 and 2 ppb, there was no depression of carbon-14 uptake.

The findings at these levels were as follows:[57]

Species	Effect of DDT on C^{14} Uptake*	
	1 ppb	2 ppb
Skeletonema costatum	—	—
Coccolithus huxleyi	+	+
Pyramimonas sp.	—	+
Peridinium trochoideum	+	+
Mixed culture**	+	+

* Increases over controls, +; decreases —.
** Typical neritic phytoplankton community.

According to these fragmentary findings, a saturated solution of DDT in sea water would not depress photosynthesis. Wurster states:

> The fact that these data apparently follow sigmoid curves is typical of dose-response relations and suggests the absence of a threshold concentration of DDT below which no effects occur. Experimental scatter produced some uptake of C^{14} above 100 percent at low concentrations of DDT, however. This should not be interpreted as a low-level stimulatory effect, a possibility that cannot be evaluated from these data.[58]

Obviously if the data show an indication, even slight, of a low-level stimulatory effect, it follows that the same data cannot be used to postulate a "no threshold" situation. Wurster pointed out that "water near a direct application of DDT to the environment, however, commonly contains concentrations comparable to those applied by me."[59] This is a far cry from stopping photosynthesis in the oceans as suggested by Ehrlich. Wurster goes on to state that "selective toxic stress by DDT on certain algae" may "favor species normally suppressed by others, producing population explosions. . . . Such effects are insidious."[60] (The word "insidious" is a favorite word in the lexicography of DDT.) Increases and decreases in aquatic photosynthesis can both be blamed on DDT. It is a well-documented observation that over-application of DDT to green crop plants may cause not only depression of photosynthesis, but death. Over-application of various chemicals kills plants; common salt, for example, used to be used as a weed-killer.

To produce a concentration of 1 ppb of DDT in the 300 million cubic miles of sea water in the oceans would take 9,000 years if the total annual production of DDT, 300 million pounds, were dispersed in the oceans each year and there was no breakdown. If the half-life of DDT in sea water is one thousand years or less, this concentration would never be reached.

The absurdity of these figures illustrates the need for quantitative examination of allegations, but is not intended as a suggestion that it is safe or desirable to use the ocean as a sink for pollutants.

Conclusion

This article has treated only a few of the points that demonstrate DDT's ultimate safety and significant contributions to man. I have chosen to discuss these few points at length, rather than to mention a large number of topics briefly. A final response, however, must be made to the charge that DDT has injurious effects on human beings. The particular charge to which I respond is based on the observation that DDT inhibits certain enzymes in laboratory experiments. Contrary to what DDT opponents would have us believe, a similar observation can be made of any of a number of substances. For example, many of the substances that we eat—salt, phosphate, magnesium, citric acid—drastically affect enzymes in test tube experiments. These experiments, moreover, do not involve intact animals. By contrast, the everyday use of DDT does involve intact animals, and the effects are remarkably minimal. In point of fact, *several hundred million people* have been exposed to DDT for prolonged periods of time without any sign of ill effect. Some of them have received heavy doses, over periods at up to 19 years[61]—the 130,000 spraymen listed by the WHO, for example, and numerous people in DDT factories and formulating plants—all without any reports of illnesses attributable to DDT. In the 1940's, enthusiastic volunteers allowed themselves to be used as experimental animals and swallowed large amounts of DDT, in some cases daily for prolonged periods.[62] For some reason, nothing seemed to happen except transient tingling of the extremities. Some day the true story of DDT, buried in the scientific literature, will be brought out into the open.

We may certainly expect an increase in dialogue between scientists and lawyers as the number of legal questions involving the environment grows apace. Such dialogue, however, when it occurs in the tightly structured setting of the courtroom, may have unfortunate consequences. If a scientist is asked, for example, whether a pesticide is poisonous, he will say "Yes;" at the same time, however, he may recommend its use at an appropriate level as a proper response to human needs. A skillful attorney can effectively exploit such an apparent contradiction, following which the scientist will probably withdraw into his shell. In the 1969 Wisconsin hearings on DDT, such incidents repeatedly took place during examinations by Environmental Defense Fund (EDF) attorney, Victor J. Yannacone, Jr. An EDF spokesman, C. F. Wurster, expressed much satisfaction with Mr. Yannacone's efforts:

Victor J. Yannacone, Jr., the Environmental Defense Fund's attorney, has an impressive grasp of scientific material, especially the environmental sciences. His cross-examination is usually aggressive and may be devastating where a witness is scientifically weak.[63]

Wurster, however, later altered his opinion. On March 9, 1971, he stated:

> It has come to my attention that certain remarks, attributed to me by Mr. Victor J. Yannacone, Jr. in May 1970, have been inserted into the record of your hearings on pesticides during the testimony of Edward Lee Rogers of the Environmental Defense Fund.
>
> I wish to deny all of the statements of Mr. Yannacone. His remarks about me, attributable to me, and about other trustees of EDF are purely fantasy and bear no resemblance to the truth. It was in part because Mr. Yannacone lost touch with reality that he was dismissed by EDF, and his remarks of May 1970 indicate that his inability to separate fact from fiction has accelerated.[64]

Yannacone's fiery and uncompromising onslaught on DDT in Wisconsin was evidently undertaken for the purposes of advocacy, for on September 27, 1970, he said, in comparatively moderate tones:

> Any law simplistically banning the use, sale, manufacture or distribution of DDT in your state, county, city or even the United States, without at the same time establishing an ecologically sophisticated pesticide regulation program, is a bad law. It won't satisfy anyone very long and will permanently polarize agriculture and conservation to such an extent that common problems can no longer be considered in rational discourse.[65]

These remarks by Mr. Yannacone reflect a scientific sophistication that he did not reveal during the Wisconsin hearings. Regrettably, however, the earlier remarks seem to have gained greater attention.

This anecdote serves to illustrate an essential difference between the advocate and the scientist. The advocate seeks a prompt and unequivocal decision on a particular issue. The scientist, however, simply cannot expect such results; he thus strives to obtain facts which enable him to ascend the spiral staircase of knowledge. With each step of his ascent, he obtains a wider and superior view of his subject matter, and, although he never reaches the top, he should on his way up help his fellow men by telling them what he sees.

[1] J. Slater, "Third World Ecology Course to Be Offered," *The Journal of Educational Change* 2:1 (Oct. 1970).

[2] Rachel L. Carson, *Silent Spring* (1962).

[3] Advertisements by National Audubon Society, 1970, in various magazines.

[4] "Our Ecological Crisis," *National Geographic Magazine* (Dec. 1970).

[5] U. S. Court of Appeals, District of Columbia, Circuit Petition for Review of Order of U. S. Dept. of Agriculture, No. 23, 813 (Apr. 6, 1971).

[6] Martin G. Garcia (WHO), Personal Communication to S. Rotrosen (June 19, 1969).

[7] See P. DeKruif, "The Magic Bullet," *Microbe Hunters*, p. 308 (1926).

[8] R. Pal, "Contributions of Insecticides to Public Health in India," *World Review of Pest Control* 1:6 (1962).

[9] P. F. Russell, The United States and Malaria: Debits and Credits *Bulletin of the New York Academy of Medicine* 44:623 (1968).

[10] *Id.*

[11] T. H. Jukes, "DDT: The Chemical of Social Change," *Clinical Toxicology* 2:359 (Dec. 1969).

[12] "Environmental Scorecard," *Richmond Independent* (Calif.) 10 (Dec. 17, 1970).

[13] E. Chaney, *Conservation News* (June 15, 1970).

[14] W. J. Hayes, Jr., "Toxicity of Pesticides to Man: Risks from Present Levels" *Proc. Roy. Soc. B.* 167:101 (1967).

[15] W. J. Hayes, Jr., W. E. Dale, and C. I. Prikle, "Evidence of Safety of Long Term, High, Oral Doses of DDT for Man," *Arch. Environ. Health* 22:119 (1971).

[16] J. M. Barnes, "Food and Health—The Safe Use of Pesticide," *British Food Journal* (May–June 1967).

[17] *Id.*

[18] E. P. Laug, F. M. Kunze and C. S. Prickett, "Occurrence of DDT in Human Fat and Milk," *Arch. Indus. Hyg.* 3:245 (1951).

[19] G. E. Quinby, J. F. Armstrong, and W. F. Durham, "DDT in Human Milk," *Nature* 207:726 (Aug. 14, 1965).

[20] Personal communication from W. J. Hayes, Jr. to P. Gyorgy, (June 4, 1970).

[21] 21 U.S.C. §348 (c) (3) (A).

[22] U. S. Department of Health, Education and Welfare, Report of the Secretary's Commission on Pesticides and Their Relation to Environmental Health, pts. I and II, at i-xvii and 1-677 (Dec. 1969).

[23] J. R. M. Innes, B. M. Uland, M. G. Valerio, L. Petrucelli, L.

Fishbein, E. R. Hart, and A. J. Pallotta; R. R. Bates, H. L. Falk, J. J. Gart, M. Klein, I. Mitchell, and J. Peters, "Bioassay of Pesticides and Industrial Chemicals for Tumorgenicity in Mice: A Preliminary Note," *J. Natl. Cancer Inst.* 44:1101 (1969).

[24] R. D. Lillie and M. I. Smith, Pathology of Experimental Poisoning in Cats, Rabbits and Rats with DDT" *U. S. Pub. Health Reports* 49:979 (1944).

[25] A. R. Cameron and F. Burgess, *British Medical Journal* 1:865 (1945).

[26] *Id.*

[27] O. G. Fitzhugh and A. A. Nelson, The Chronic Oral Toxicity of DDT, *J. Pharmacol.* 89: 18 (1947).

[28] *Id.*

[29] E. P. Laug, A. A. Nelson, O. G. Fitzhugh, and F. M. Kunzer "Liver Cell Alteration and DDT Storage in the Fat of the Rat Induced by Dietary Levels of 1 to 50 ppm DDT," *J. Pharmacol.* 98:268 (1950).

[30] G. R. Cameron and M. B. Cheng, "Failure of Oral DDT to Induce Toxic Changes in Rats," *British Medical Journal* 2:819 (1951).

[31] A. J. Lehman, Summaries of Pesticide Toxicity (Assoc. Food and Drug Officials of the U. S., 1965).

[32] W. J. Hayes, Jr. testimony at State of Washington Hearings on DDT (Seattle Oct. 15, 1969).

[33] J. R. M. Innes *et al., supra* note 23.

[34] *Id.*

[35] W. J. Hayes, Jr., *supra* note 32.

[36] The Place of DDT in Operations Against Malaria and Other Vector Borne Diseases, WHO Statement EB 47/WP/14 (Jan. 22, 1971).

[37] Statement by Committee on Experimental Toxicology, *J. Am. Med. Assoc.* 212:1055 (1970).

[38] *Id.*

[39] National Communicable Disease Center, Department of Health, Education and Welfare, DDT in Malaria Control and Eradication (July 12, 1969).

[40] Sir Macfarlane Burnet, *Natural History of Infectious Disease,* (2d ed. 1953).

[41] R. Pal, *supra* note 8.

[42] "DDT Killing Crabs," *San Francisco Chronicle* (May 3, 1969).

[43] "What Happened to Our Crabs? Pollution!" *San Francisco Examiner,* p. 11. (Nov. 14, 1970).

[44] *Id.*

[45] *Science*, Vol. 162 (Feb. 7, 1969).

[46] *San Francisco Chronicle* (June 13, 1970).

[47] "Science for the Citizen," *Scientific American* p. 86 (Sep. 1970).

48 J. D. Lockie, D. A. Ratcliffe, & R. Balharry, *Journal of Applied Ecology* 6:381 (1969).

49 J. J. Hickey, "Eastern Population of the Duck Hawk," *Auk* 59:176 (1942).

50 R. A. Herbert and K. G. S. Herbert, "The Extirpation of the Hudson River Peregrine Falcon Population," *Peregrine Falcon Populations—Their Biology and Decline* pp. 133–154 (1959).

51 R. Fyfe, The Peregrine Falcon in Northern Canada," *Peregrine Falcon Populations—Their Biology and Decline*, pp. 101–114 (1959).

52 J. H. Enderson & D.D. Berger, *Condor* 70:149 (1968).

53 D. A. Ratcliffe, "The Peregrine Situation in Great Britain," *1965–66 Bird Study* 14, 238 (1968).

54 Advisory Committee on Pesticides and Other Toxic Chemicals, Department of Education and Science, London, *Further Review of Certain Persistent Organochlorine Pesticides Used in Great Britain* (1969).

55 P. Ehrlich, "Eco-catastrophe," *Ramparts* 8:24 (Sep. 1969).

56 C. F. Wurster, "DDT Reduces Photosynthesis by Marine Phytoplankton," *Science*, 159:1474 (1968).

57 *Id.*

58 *Id.*

59 *Id.*

60 *Id.*

61 W. J. Hayes, Jr., W. E. Dale, and C. I. Prinkle, "Evidence of Safety of Long-Term High, Oral Doses of DDT for Man," *Arch. Environ. Health*, 22:119 (1971).

62 W. J. Hayes, Jr., in "DDT The Insecticide Dichlorodiphenyltrichloroethane and its Significance," *Human and Veterinary Medicine* 11:252 (P. Muller, ed., 1959).

63 C. F. Wurster, "DDT Goes to Trial in Madison," *BioScience* 19:809 (1969).

64 C. F. Wurster, in transcript of Congressional Hearings on Agriculture, Representative Poage, Chairman, (Mar. 9, 1971).

65 V. J. Yannacone, Jr., in *Highlights 70*, Congress for Recreation and Parks, Philadelphia, Pa., p. 24, (Sep. 27–30, 1970).

Exposure of
Formulating Plant Workers to DDT

Homer R. Wolfe, and John F. Armstrong

SEVERAL previous studies of exposure of workers to DDT in manufacturing and formulating plants have been carried out in this and other laboratories.[1-5] In these studies, indirect methods for measurement of exposure were employed which involved determination of levels of DDT and DDT-derived material in blood, fat and other tissues, and excretion of DDA in the urine as an indication of exposure of workers to DDT. The measurements reflect DDT absorption into the body. The workers studied had varying degrees of protection from exposure, ranging from the use of very little protective clothing to relatively adequate protection, including the wearing of respirators.

Ortelee,[1] in an exposure study of workers in manufacturing and formulating plants, estimated DDT absorption by comparing urinary DDA concentrations with those obtained in a study[6] of human volunteers on known daily oral dosages of technical DDT.

Estimation of potential exposure was classed as "slight," "moderate," or "heavy," based on observations of the men at their jobs and also on an estimate by the men and by their superiors as to the degree of their exposure. In another study of exposure of workers in manufacturing and formulating plants, Laws et al[2] estimated that the average daily intake of DDT by 20 men with high occupational exposure was 17.5 to 18 mg per man per day as compared with an average of 0.04 mg per man per day for the general population. These figures were based on storage of DDT in fat and excretion of DDA in urine.

Edmundson et al[4,5] also utilized indirect methods of measurement of exposure of pesticide formulators to DDT by determining levels of DDT and DDE in blood and of DDA in the urine, as did Durham et al[3] in determining DDA excretion levels in workers, including those exposed in formulating plants.

Although direct methods for measurement of potential exposure to DDT have been carried out for workers applying pesticides,[7-10] we are not aware of any studies of direct measurements of potential exposure of formulating plant workers to DDT. The present studies, using direct methods, were conducted to determine the amount of impingement of DDT on body surfaces and the amount of DDT entrapped on respirator pads during given periods of work in formulating plants. The values obtained were used to estimate the amount of DDT the workers were subjected to as potential exposure when using certain recommended protective gear or clothing, and also to estimate the maximum potential exposure of workers when no special protective clothing or devices are used. DDA excretion was determined in several instances for correlation with potential exposure data.

Materials and Methods

The study was conducted in two plants formulating 50% water-wettable DDT powder. In order to determine the potential exposure in different work situations, exposure pad studies were carried out on (1) the worker who inserted the proper proportions of ingredients into the formulating machine ("mixer"), (2) the worker who filled bags with pesticide at the filler spout ("bagger"), (3) the worker who operated a machine to sew tops of bags closed ("sewer"), and (4) the worker who packed filled bags into cartons ready for shipment ("packer"). Exposure tests were carried out during work periods lasting from 30 minutes to 1 hour.

The amount of DDT to which a worker potentially would be subjected during work activities was estimated by the techniques and procedures described by Durham and Wolfe.[11] Dermal contamination was measured primarily by attaching layered-gauze absorbent pads to various parts of the worker's body or clothing and allowing them to be exposed during a timed period of work. Respiratory exposure was estimated from the contamination of special filter pads used in place of the usual outer absorbent filter pads which cover the filter cartridges of the respirators worn by the subjects. The filter pads were covered with plastic funnels modified to a specific aperture size to reproduce as nearly as possible the aerodynamics of air flow through the nostrils. The funnels also prevented direct impingement of particles onto the pad except for those carried through the apertures by respiratory action. This technique renders it unnecessary to measure total air volume, because all inhaled air passes through the filter pads.

Respirator and dermal exposure pads were extracted with benzene in a Soxhlet apparatus, and samples were analyzed for DDT by electron-capture gas-liquid chromatography. Values obtained were used to calculate the milligrams of potential exposure per man per hour of work. The percent of the toxic dose to which workers were potentially exposed was calculated by the procedure described by Durham and Wolfe,[11] based on comparison between the dermal and respiratory exposure values determined in the present study and animal toxicity figures published by Gaines.[12]

Single urine samples were collected from 17 different workers following extended DDT formulation operations in the plants. In addition, two workers who seemed to be receiving the heaviest exposure collected complete 24-hour specimens of urine for a period of five days so that the rate of DDA excretion could be determined. A total of 171 urine samples were analyzed for DDA content by the method of Cueto et al.[13] The DDA was expressed as DDT equivalent (actual DDA × 1.24).

Results and Comment

Potential Dermal and Respiratory Exposure.—Table 1 and 2 present exposure values for work stations in two formulating plants. The calculations in Table 1 are not based on the actual protective gear worn but on the use of minimum protection (no respirator, shirt with short sleeves, open collar, no hat, no gloves; it was assumed that the clothing worn gave complete protection of areas covered). The purpose of this was to arrive at values that might reflect the maximum potential exposure that could occur in different work situations where proper protective gear was not utilized. Table 2 shows values for the same workers but calculations were based on actual protective gear worn by the different individuals.

In plant A, no values were obtained for the mixer, and in plant B no values were obtained for the packer; no sewer was employed in plant B because of the use of a bag sealing machine. However, it is believed that the work stations studied were a fair representation of the plant working areas which have the greatest potential for pesticide contamination. Observations indicated that the potential exposure would be much higher in plant A than in plant B, not only because the ventilation system was obviously inadequate in plant A but also because the presence of pesticide dust on the equipment, floor, and building structural members indicated a laxity in plant housekeeping. These observations were confirmed by the results.

As can be seen in Table 1, the highest potential exposure values were for plant A. An example of the difference in potential exposure between the two formulating plants can be seen in values for the workman (bagger) who filled 4- and 5-lb bags with pesticide. The bagger in plant A would have been subjected to a potential dermal exposure of from 95.0 to 992.9 mg/hr, with a mean of 524.5 mg/hr of DDT had he not worn recommended protective gear. This is approximately 17 times more potential dermal exposure than for the bagger in plant B (range, 24.6 to 34.0 mg/hr; mean, 31.2), where more adequate ventilation and better housekeeping practices were in force. The potential respiratory exposure value obtained for this work position in plant A was from 4.20 to 33.80 mg/hr, with a mean of 14.11 mg/hr, or approximately 88 times more than the 0.14 to 0.19 mg/hr (mean, 0.16 mg/hr) in plant B. Such great variation in exposure conditions in formulating plants is not uncommon.

Some differences in potential exposure can be noted for the different work positions. In plant A, the worker at the 4- and 5-lb bagging station was subjected to the highest exposure, followed by the sewer, the bagger at the 50-lb bagging station, and then the packer. Inadequate ventilation at the filler spout appeared to contribute most importantly to exposure of the worker who filled the 4- and 5-lb bags. Occasionally, malfunction of the filler spout mechanism resulted in gross contamination of the worker. The authors have observed that these two problems are not uncommon in pesticide formulating plants. During one work period, malfunction of the filler spout was responsible for the very high respiratory value of 33.8 mg/hr for the bagger. The sewer station was located only 5 feet from the bagging operation, and it was observed

that contamination from the bagging station was reaching the sewing machine area. The packer who packed bags of pesticide into cardboard shipping cartons was located approximately 15 feet from the bagging station and received very little contamination from that area. At the packing station, the main exposure appeared to be from small amounts of water-wettable DDT powder being blown from inside the carton onto the face and upper body area as the worker pressed the filled bags tightly in place.

The 50-lb bagging station was located in a separate area and required no sewing or packing operation. At this location, much of the exposure of workers appeared to be from contact with dry pesticide that had been blown into the air as filled bags were removed from the filler spout and again when the flap over the filler hole of the bag was being closed by folding.

If all the DDT from the potential dermal and respiratory exposure to which a worker might be subjected when wearing a minimum of protective gear (as shown in Table 1) were completely absorbed, the calculated percent of toxic dose per hour of exposure (highest value of 0.76% for the bagger in plant A) is not particularly high as far as potential for producing acute toxic effects is concerned. However, possible long-term effects of excess absorption and storage in the body resulting from continued high exposure to DDT should be a subject of greater concern than the more remote possibility of acute toxic effects to workers. Thus, the relatively high potential dermal and respiratory exposure to this compound in formulating plants, even though DDT is of much lower toxicity to warm-blooded animals than many other compounds in common use, emphasizes the importance of following safety precautions to prevent excess exposure.

Exposure values in Table 2 are presented as an estimation of potential exposure in the different work stations, taking into account the protective clothing and respirators actually worn by the workers. Workers in both formulating plants wore rubber boots, cap, respirator, cloth coveralls with long sleeves but with top button at neck open, and rubber gauntlet gloves, which provided some protection at the wrist area not covered by

Table 1.—Potential Exposure of Formulating Plant Workers to DDT, Based on Minimum Protection*

Work Position†	Size of Bags	No. of Work Periods Studied		Dermal (mg/man/hr)	Respiratory (mg/man/hr)	Total (Percent Toxic Dose per Hour)
Plant A						
Bagger	4 & 5 lb	4	Range	95.0-992.9	4.20-33.80	0.08-0.76
			Mean	524.5	14.11	0.39
Bagger	50 lb	4	Range	105.8-226.5	3.26-11.80	0.09-0.20
			Mean	153.5	6.07	0.12
Sewer	4 & 5 lb	4	Range	32.6-706.8	0.42-9.05	0.03-0.45
			Mean	498.0	4.86	0.31
Packer	4 & 5 lb	4	Range	69.2-146.7	0.91-7.22	0.005-0.13
			Mean	98.5	4.00	0.08
Plant B						
Bagger	4 lb	5	Range	24.6-34.0	0.14-0.19	0.015-0.020
			Mean	31.2	0.16	0.019
Mixer	4 lb	5	Range	13.3-44.7	0.11-0.61	0.014-0.030
			Mean	32.7	0.26	0.020

* Minimum protection: no respirator, shirt with short sleeves and open collar, no hat or protective gloves, leather shoes; it is assumed that clothing worn gave complete protection of areas covered.
† Bagger, worker who fills bags with pesticide at the filler spout; sewer, worker who operates a machine to sew tops of bags closed; packer, worker who packs filled bags into cartons ready for shipment; mixer, worker who inserts the proper proportions of ingredients into the formulation machine.

Table 2.—Potential Exposure of Formulating Plant Workers to DDT, Based on the Wearing of Certain Protective Clothing and Devices*

Work Position†	Size of Bags	No. of Work Periods Studied		Dermal (mg/man/hr)	Respiratory (mg/man/hr)	Total (Percent Toxic Dose per Hour)
Plant A						
Bagger	4 & 5 lb	4	Range	26.4-124.4	0.083-0.676	0.015-0.075
			Mean	72.7	0.282	0.043
Bagger	50 lb	4	Range	24.3-87.5	0.065-0.236	0.014-0.051
			Mean	43.5	0.122	0.025
Sewer	4 & 5 lb	4	Range	10.7-223.2	0.008-0.181	0.006-0.130
			Mean	71.2	0.097	0.040
Packer	4 & 5 lb	4	Range	15.2-32.5	0.018-0.144	0.009-0.019
			Mean	26.2	0.080	0.015
Plant B						
Bagger	4 lb	5	Range	7.8-19.6	0.002-0.004	0.004-0.011
			Mean	14.3	0.003	0.008
Mixer	4 lb	5	Range	4.2-18.2	0.003-0.012	0.002-0.010
			Mean	12.9	0.005	0.007

* Coveralls with long sleeves and open collar, cap, rubber gauntlet gloves; it is assumed that the clothing worn gave complete protection of areas covered and respirators worn provided 98% respiratory protection.
† Bagger, worker who fills bags with pesticide at the filler spout; sewer, worker who operates a machine to sew tops of bags closed; packer, worker who packs filled bags into cartons ready for shipment; mixer, worker who inserts the proper proportions of ingredients into the formulation machine.

	Worker No.	DDA Levels Expressed as DDT Equivalent (ppm)	
		Day 1	Day 2
Plant A: areas of greatest contamination*			
	1	0.04	0.10
	2	0.68	...
	3	0.18	0.18
	4	0.57	0.14
	5	0.50	0.50
	6	0.16	0.14
	7	0.27	0.38
	8	0.26	0.27
Plant A: areas of less contamination†			
	9	0.04	0.02
	10	0.14	0.12
	11	...	0.07
	12	0.02	0.02
	13	...	0.35
	14	0.40	0.17
	15	...	0.61
Plant B: areas of greatest contamination*			
	16	0.08	0.07
	17	0.06	0.04

* The general areas of bagging, sewing, mixing, or packing operations.
† Areas of less contamination such as storage, shipping, or other areas. Workers 13 and 14 were foremen who moved to all areas.

the sleeve. In plant A, the caps worn were the type with a bill at the front; the bill was considered to give some protection from downward-moving particles in the face-front area but very little protection on other face-neck areas. In plant B, the caps worn were the beanie type with no bill, and provided protection only for the top of the head. Also, rubber aprons were worn in plant B. Calculations of potential exposure were based on the assumptions that the clothing worn gave complete protection of the body areas covered and that the respirators fit tightly enough to allow passage of no more than 2% of the potential respiratory exposure. Exposure values were calculated for each worker, taking into account the estimated percentage of skin area actually exposed.

Data in Table 2 indicate that wearing additional protective gear reduces the potential exposure considerably. By comparing the mean values for percent of toxic dose received per hour with those calculated in Table 1, it can be seen that had the workers not worn the recommended added protective clothing and respirators they would have been subjected to approximately five to ten times more DDT in plant A and approximately 2½ to 3 times more in plant B.

Unfortunately, the wearing of clothing which provides only minimum protection and nonuse of a respirator are not particularly uncommon in formulating plants where compounds of relatively low toxicity, including DDT, are being handled. Ortelee,[1] in his study of workers in manufacturing or formulating plants, observed that in most cases no special protective clothing or respirators were used by the men while working with DDT. Thus, it is not surprising that his calculations indicated that the workers had absorbed an average of about 200 times as much DDT as that absorbed by the general population from dietary sources. Our observations in formulating plants indicate that, when workers are at the bagging station and not wearing respirators, it is almost impossible for them to avoid inhaling large quantities of the compound which is sometimes obviously present in high concentrations near the breathing zone. When 50% water-wettable DDT powder is involved, the intake may be quite significant. In view of the occurrence of occasional potential respiratory exposure as high as 33.8 mg/hr, as reported in the present study, or even much less than that amount, the 647 ppm of DDT-derived material found in the fat of a

45

Excretion of DDA (DDT equivalent) by a formulating plant worker following several exposure periods. Sample period represents interval covered by collection of a voiding of urine (total number of hours per voided specimen).

46

formulating plant worker by Laws et al[2] is not difficult to comprehend.

Urinary Excretion of DDA.—Levels of the metabolite DDA (expressed as DDT equivalent) found in urine of workers are shown in Table 3. Single urine samples were collected on two different days from men in different work areas. DDA in the urine of workers who spent most of their work day in the bagging, sewing, mixing, or packing areas generally was higher than that found in the urine of workers spending more time in areas of less DDT contamination. This indicates that even though workers in the areas of higher contamination wore more protective gear they absorbed more DDT than the other workers. The lack of high values for the two workers in plant B reflects the lower estimated potential dermal and respiratory exposure for these individuals. It was not possible to establish definite excretion values for specific work stations because of the occasional shifting of personnel from one work position to another.

Workmen listed as numbers 13 and 14 were foremen and moved to all areas of the plant. The level of DDA in urine of these two individuals was in the range of that of workers in areas of greatest contamination, probably because the foremen did not wear respirators or any added protective clothing when spending limited periods of time in such areas.

In addition to single urine specimens on most workers, two workers who appeared to be receiving the heaviest exposure collected complete 24-hour urine specimens for a period of approximately five days. Each voiding was collected in a separate container so that the rate of DDA excretion could be determined. The accompanying Figure shows excretion of DDA (as DDT equivalent) for one of these subjects. Although this individual excreted slightly more than the other worker, the excretion levels in relation to exposure were similar; therefore, the curve showing amount excreted per hour is given for only the subject having the greatest excretion. This worker served as a packer during portions of the first and third days, and as a bagger during portions of the second, fourth, and fifth days. On the sixth day, Saturday, the subject did not work; and, thus, no exposure to DDT was involved.

The excretion levels correlated quite well with exposure. Following each exposure period, the excretory level of DDA increased and reached a maximum, on the average, 10.6 hours after exposure began. This value agrees closely with the maximum excretion time of 10.1 hours found by Wolfe et al[14] for pesticide applicators exposed to liquid DDT sprays in the field, and is near the 14 hours reported by Edmundson et al[5] for workers exposed to DDT in formulating plants. The mean maximum excretion rate for the five days was 12.6μg/hr, with a range of 6.5μg to 16.9μg/hr. The total excretion of DDA during the period was only 998.9μg.

As can be seen in the Figure, DDA excretion fell to a very low level during the sixth day, which was not a work day. The subject reported that he remained relatively inactive inside his dwelling most of the day. Thus, he was not only less active but he was also not subjected to the higher outside temperatures. These factors, in addition to the termination of DDT exposure, may have contributed to the marked rate of decrease in excretion.

Although at the time of this writing the future use of DDT appears somewhat uncertain, it seems proper to add to the record of this compound its potential for dermal and respiratory contamination of workers in certain formulating plants. The finding of relatively low levels of DDA excreted, as compared with the relatively high potential exposure, indicates that the protective gear used by the workers was effective. If protective gear had not been worn there undoubtedly would have been more excretion of DDA, as was found by Ortelee[1] and Laws et al[2] in their studies of formulating plant workers. Even though all but the high values for DDA content of urine of workers in the present study fall within the range of levels excreted by the general population reported by Durham et al,[3] correlation of excretion levels with exposure, seen in the Figure, is evidence that increased excretion was due to exposure in the formulating plant.

The finding of relatively high values for potential exposure to DDT and relatively low DDA excretion in urine also indicates that there was a minimum of dermal absorption of DDT at unprotected skin areas. According to lethal dose for 50% survival of the group values reported by Gaines,[12] technical DDT is less toxic dermally to white rats than was a large percentage of the many pesticide compounds he tested. Draize et al,[15] following a study of percutaneous absorption of DDT, concluded that DDT in the dry or powdered form is poorly absorbed by the skin. Although the health status of workers in this study was not specifically examined, the fact that they did not report any illness related to their DDT exposure correlates well with low human toxicity of DDT from other studies.[1,2,6]

Even though the absorption of DDT through the skin is not as great as with certain other pesticides, such as organophosphorus compounds, it seems advisable to use protective clothing to cover as much of the skin area as possible in order to avoid absorption and storage of the compound in the body. Where exposure is to dry formulations of the compound, the respiratory route takes on increased significance and warrants the use of tight-fitting filter cartridge-type respirators when working in areas of greater contamination within the formulating plant. When a great reduction of potential exposure can be brought about with the use of special protective gear, as shown in this study, it seems negligent for any employer to allow less protection even if a compound has a low toxicity.

Dorothy Fillman assisted with the insecticide determinations.

References

1. Ortelee MF: Study of men with prolonged intensive occupational exposure to DDT. *Arch Industr Health* 18:433-440, 1958.

2. Laws ER Jr, Curley A, Biros FJ: Men with intensive occupational exposure to DDT: A clinical and chemical study. *Arch Environ Health* 15:766-775, 1967.

3. Durham WF, Armstrong JF, Quinby GE: DDA excretion levels: Studies in persons with different degrees of exposure to DDT. *Arch Environ Health* 11:76-79, 1965.

4. Edmundson WF, Davies JE, Nachman CA, et al: *p,p'*-DDT and *p,p'*-DDE in blood samples of occupationally exposed workers. *Public Health Rep* 84:53-58, 1969.

5. Edmundson WF, Davies JE, Cranmer M, et al: Levels of DDT and DDE in blood and DDA in urine following a single intensive exposure. *Industr Med Surg* 38:145-150, 1969.

6. Hayes WJ Jr, Durham WF, Cueto C Jr: Effect of known repeated oral doses of chlorophenothane (DDT) in man. *JAMA* 162:890-897, 1956.

7. Hayes WJ Jr: Pharmacology and toxicology of DDT, in Müller P (ed): *DDT: The Insecticide Dichlorodiphenyltrichloroethane and Its Significance.* Basel, Switzerland, Birkhäuser Verlag, 1959, vol 2, pp 9-247.

8. Wassermann M, Iliescu S, Mandric G, et al: Toxic hazards during DDT- and BHC-spraying of forests against *Lymantria monacha. Arch Industr Health* 21:503-508, 1960.

9. Wolfe HR, Durham WF, Armstrong JF: Exposure of workers to pesticides. *Arch Environ Health* 14:622-633, 1967.

10. Wolfe HR, Walker KC, Elliott JW, et al: Evaluation of the health hazards involved in house spraying with DDT. *Bull WHO* 20:1-14, 1959.

11. Durham WF, Wolfe HR: Measurement of the exposure of workers to pesticides. *Bull WHO* 26:75-91, 1962.

12. Gaines TB: Acute toxicity of pesticides. *Toxic Appl Pharmacol* 14:515-534, 1969.

13. Cueto C, Jr, Barnes AG, Mattson AM: DDT in humans and animals: Determination of DDA in urine using an ion exchange resin. *J Agric Food Chem* 4:943-945, 1956.

14. Wolfe HR, Durham WF, Armstrong JF: Urinary excretion of insecticide metabolites: Excretion of para-nitrophenol and DDA as indicators of exposure to parathion. *Arch Environ Health* 21:711-716, 1970.

15. Draize JH, Nelson AA, Calvery HO: The percutaneous absorption of DDT (2,2-bis(p-chlorophenyl) 1,1,1-trichloroethane) in laboratory animals. *J Pharmacol Exp Ther* 82:159-166, 1944.

Chlorinated Hydrocarbon Pesticide Residue in Human Tissues

Donald P. Morgan, MD, PhD, and
Clifford C. Roan, PhD

PERSISTENT PESTICIDES are of concern when their residues possess—in addition to persistence—toxicity, mobility in the environment, and a tendency for storage in the biota. Although current evidence does not indicate that present concentrations of pesticide in man's food and environment produce an adverse effect on his health, there continues to be a general public concern with the relationship between man and the pesticide residues in his body tissues and fluids. Specific biologically important questions that arise are the following: (1) the influence of local pesticide usage and food residues on levels in human tissue, (2) long-term trends in human tissue storage, (3) relationships between disease states and tissue pesticide concentrations, and (4) the distribution of pesticides in tissues and organs of the human body.

These questions will be answered when analytical precision substantially exceeds the measurement differences to be evaluated. Although the analytical arts and sciences have yet to achieve ultimate precision, gas-liquid chromatography is beginning to supply partial answers to these four questions.

Pesticide analysis of autopsy tissues has been suggested as one approach to these areas of interest. Although our study of 70 tissue sets from autopsies performed in Tucson, Ariz, in 1966 to 1968 has not yielded definitive answers, the experience has pointed out potentialities and limitations of this approach to the problems of human pesticide storage.

Methods

Necropsy Material.—Nearly all of the tissue sets included adipose tissue, liver, and kidney. Specimens of brain (cerebrum) were available in 52 of them and whole blood (heart chamber) samples accompanied 37 of these sets. Approximately half of the cases were the result of sudden death by reason of accident or medical catastrophe, the remainder of chronic disease. Most of the tissues were embalmed and kept in frozen storage prior to analysis.

Pesticide Analysis.—Tissue extraction and cleanup were performed basically according to the methods of Mills et al[1,2] with several modifications.

Six grams of liver, kidney, brain, or whole blood were homogenized for five minutes with 30 to 35 ml of a mixture of ethanol, ethyl ether, and hexane (1:1:1, percent volume in volume). Contents were then transferred quantitatively to a separatory funnel containing 250 ml of distilled water. After two inversions, the lower aqueous phase was discarded, and the solvent phase was washed twice again with distilled water. The extract was then evaporated over a water bath to 2 to 3 ml, transferred to a centrifuge tube and diluted to exactly 12 ml with hexane. Ten milliliters of the extract were then passed through a florisil column (60/100 mesh, 20 gm), previously partially deactivated with 2 ml distilled water. Elution was accomplished with 200 ml of a mixture of ethanol, ethyl ether, and hexane (0.6:29.4:170, v/v). The eluate was evaporated under a stream of filtered dry air to suitable volume for gas-liquid chromatographic analysis.

Specimens of adipose tissue and very fatty livers were limited to 1.25 gm. The primary ethanol-ethyl ether-hexane extract was partitioned (1:1, v/v) with hexane-saturated acetonitrile four times, serially. After adding 10 ml of hexane to the separated acetonitrile fraction, the acetonitrile was washed out by three serial additions of 300 to 400 ml of distilled water. The remaining hexane was then subjected to column cleanup as described.

Samples were analyzed on a chromatograph with tritium foil electron capture detection. Operating specifications for the instrument were as follows: column, Pyrex 6 ft × 1/8 inch (internal diameter); column packing, chromatographic support 60/80 mesh; and column coating, 10% Dow 11 and 15% QF-1. The inlet temperature was 230 C; detector temperature, 195 C; column temperature, 205 C; and nitrogen flow, 20 ml/min. The polarizing voltage was 12 and the quantitation was made by peak height measurements of the electron capture chromatograms. Most but not all measurements were confirmed by microcoulometric detection.

Tissue Lipid Measurements.—Two milliliters of the 12 ml hexane extract of tissue sample were evaporated to dryness at room temperature. The residue was accurately weighed and related to the whole tissue weight from which it was derived.

Findings

Geographic Differences in Human Tissue Pesticide Storage.—Table 1 shows average DDT (2,2,bis[p-chlorophenyl]1,1,1-trichloroethane), DDE (2,2,bis[p-chlorophenyl]1,1-

Table 1.—Whole Tissue Pesticide Concentrations (ppm) Determined by Gas Liquid Chromatography in Tucson, Hawaii and Miami, 1966 to 1968*

	Adipose			Liver		
	Tucson	Hawaii†	Miami	Tucson	Hawaii	Miami
DDT	1.54	1.30	2.81	0.101	0.047	0.120
DDE	4.58	4.51	6.67	0.432	0.200	0.350
Dieldrin	0.140	0.040	0.215	0.031	0.004	0.035
	Kidney			Brain		
	Tucson	Hawaii	Miami	Tucson	Hawaii	Miami
DDT	0.030	0.083	0.036	0.022	0.011	Trace
DDE	0.133	0.209	0.077	0.084	0.083	0.123
Dieldrin	0.011	0.006	0.013	0.007	0.003	0.035

*Data for Hawaii and Miami from the reports of Casarett et al[3] and Fiserova-Bergerova et al.[4]
† Averages of samples from three anatomic sites.

dichloroethylene) and dieldrin (1,2,3,4,10,10-hexachloro-6,7-epoxy 1,4,4a,5,6,7,8,8a-octahydro-1,4-endo, exo-5,8-dimethano-naphthalene) levels in fat, liver, kidney, and brain specimens obtained and analyzed in Tucson, Hawaii,[3] and Miami[4] in the period 1966 to 1968.

To what extent the apparent geographic differences reflect biologic reality is unknown, but experience leads us to believe that subtle variations in tissue pesticide analytical technique could account for any of the differences recorded here. By contrast, the very different adipose tissue levels of DDT and DDE recently reported from Holland[5] (20% to 40% of US values) probably do reflect the much more limited use of DDT in western Europe than in the United States. Tissue contents of dieldrin, on the other hand, are quite similar in the two countries, and this, again, corresponds roughly to what is known of usage.

Trends in Tissue Pesticide Storage in Arizona Residents During the Last Five Years.—It was concluded in 1966,[6] and confirmed in 1967,[7] that storage of pesticides in the adipose tissue of US residents had not changed materially since tissue measure-

Table 2.—Mean Adipose Tissue Levels of Chlorinated Hydrocarbon Pesticides in Arizona Residents, 1963 and 1968*

No. of Samples	Region	Year	DDT	DDE	Dieldrin
5	Phoenix	1963	2.05	5.79	0.21
70	Tucson	1968	1.54	4.58	0.14

* Data for 1963 from Dale and Quinby.[8]

Table 3.—_Comparison of Tissue Pesticide Levels in Victims of "Chronic Disease" Compared to Those Meeting "Sudden Death"_

	DDT		DDE		Dieldrin	
	Sudden Death	Chronic Disease	Sudden Death	Chronic Disease	Sudden Death	Chronic Disease
	Based on Arithmetic Means					
Adipose, ppm	1.43	1.65	4.68	4.42	0.13	0.15
Blood, ppb	7	11	33	28	4	4
Liver, ppb	114	89	506	357	47	16
Kidney, ppb	31	30	135	132	14	8
Brain, ppb	20	23	83	85	7	6
	Based on Median Values					
Adipose, ppm	1.05	0.92	4.32	2.55	0.11	0.11
Blood, ppb	5	9	21	19	4	3
Liver, ppb	45	43	238	209	14	7
Kidney, ppb	20	14	79	56	10	7
Brain, ppb	14	13	83	48	6	6

ments were first made in 1950. Our own data from southern Arizona residents tend to bear out a temporal consistency of tissue concentrations. Table 2 compares our values for adipose DDT, DDE, and dieldrin with measurements made microcoulometrically in 1963[8] on samples from the Phoenix area.

The paucity of 1963 measurements, and the geographical discrepancy compromise the validity of that comparison. However, it gives some assurance that human tissue stores of these chemicals are not increasing steadily.

Tissue Pesticide Levels in the Chronically Ill.—Table 3 shows arithmetic mean and median values of tissue pesticides from "chronic disease" and "sudden death" cases.

The disparity between means and medians indicates the skewness of some data distributions, usually attributable to a few outlying (relatively high) values.

None of the differences between "chronic disease" and "sudden death" means or medians have convincing statistical significance. If anything, "chronic disease" values seem to be slightly lower. Ranking of cases

Table 4.—_Intratissue Correlations of Pesticide Measurements With Lipid Concentrations_

Tissue	No. of Specimens	DDT	DDE	Dieldrin
Adipose	70	0.10	0.32*	0.03
Blood	37	0.28	0.81*	0.19
Liver	70	0.42*	0.46*	0.15
Kidney	66	0.77*	0.68*	0.46*
Brain	52	0.10	0.23	−0.01

* Significant at $P<0.01$.

by magnitude of pesticide concentrations, as done by Casarett et al,[3] yields no remarkable clustering by diagnosis or pathogenesis at either end of the scale.

The Relationship of Tissue Pesticide Concentration to Lipid Content.—The tendency for tissue pesticide storage to increase with lipid content is shown in the Figure, in which average whole tissue contents of DDT, DDE, and dieldrin are plotted against mean total extractable lipid in five tissues. The approximate parallelism of DDT and DDE (on logarithmic paper) indicates a fairly stable proportionality between these chemicals. The flatter curve for dieldrin implies a basically different tissue distribution. With adipose tissue as a reference, DDT and DDE levels in blood, liver, and kidney can be estimated on the basis of their respective lipid contents. These estimates agree well with the actually measured whole-tissue concentrations. In the case of dieldrin, however, measured blood and parenchymal tissue values are two to four times higher than those estimated from tissue lipid content. The inference that dieldrin is distributed extensively to nonlipid tissue components agrees well with previous studies bearing on the transport of this chemical in blood.[9]

The much lower pesticide values in brain tissue (which contains a variety of lipid materials, but virtually no neutral fat) may reflect differences in solubility, chemical affinity, tissue penetration, or combinations of these factors. Interestingly, each brain pesticide concentration corresponds to values that would be expected in a tissue of about 1% total extractable lipid, rather than 8.9% (Figure).

Specimens of a given tissue may have differing lipid contents, sometimes due to metabolic or degenerative changes (liver and blood), sometimes due to nonhomogeneity of the tissue substance (brain, kidney, adipose). Table 4 shows correlation coefficients relating DDT, DDE, and dieldrin levels to lipid concentrations in multiple specimens of each of five tissues. Individual variation in dietary intake of the three pesticides prior to death no doubt detracted from each correlation, more in the case of some chemicals than others. For DDE, the frequency of significant correlations with lipid content is impressive. For some tissues (especially brain), the range of variation of tissue lipid is small, and the likelihood of a significant correlation with pesticide concentration is correspondingly reduced.

It is reasonable to suppose that after protracted exposure to a specific pesticide, a steady state equilibrium is arrived at among the tissues of the body. Proportionality constants have been proposed by de Vlieger et al,[5] with which our own data are in rough agreement. We did not, however, discover the very high intertissue correlations reported by de Vlieger (our data, Table 5), with

Table 5.—Correlation Matrices Relating Pesticide Residues in Five Tissues*

	Adipose	Blood	Liver	Kidney	Brain
DDT					
Adipose	C	0.29	0.33‡	0.62‡	0.42‡
Blood	**0.27**	C	0.02	0.55‡	0.42‡
Liver	**0.36‡**	**0.19**	C	0.19	0.12
Kidney	**0.07**	**0.17**	**0.00**	C	0.38‡
Brain	**0.31†**	**0.31**	**0.13**	**0.10**	C
DDE					
Adipose	C	0.46‡	0.36‡	0.32‡	0.56‡
Blood	**0.38†**	C	0.36‡	0.82‡	0.66‡
Liver	**0.22**	**0.36**	C	0.18	0.37†
Kidney	**0.18**	**0.12**	**−0.03**	C	0.65‡
Brain	**0.60**	**0.23**	**0.33†**	**0.22**	C
Dieldrin					
Adipose	C	0.67‡	0.02	0.35‡	−0.03
Blood	**0.23**	C	**−0.07**	0.20	−0.02
Liver	**0.07**	**−0.04**	C	0.24	0.10
Kidney	**0.33‡**	**0.45‡**	**0.21**	C	0.24
Brain	**0.01**	**−0.24**	**0.04**	**0.26**	C

* Lightface values indicate concentration in extractable lipid; boldface, in whole tissue; C indicates a correlation coefficient of 1.00.
‡ Significant at **P** < 0.01.
† Significant at **P** < 0.05.

either "whole tissue" or "lipid extractable" as bases for measurement.

Comment

In comparison of our own data with that from other laboratories, there is little evidence of geographic differences in human pesticide stores in residents of three widely separated states. This conforms with the view that tissue pesticide comes mainly from diet and American meals are usually made up of foodstuffs gathered from the breadth of the land. Lower DDT tissue pesticide contents in Western European residents have been noted before,[10] and correlate well with low dietary residues.[11]

Changes in tissue stores, with time, are likely to be subtle, demanding precise and standardized measurement for detection. We hope that present-day methods of analysis will enjoy continued use for decades into the future, as human tissue stores change in response to changing patterns of chemical contamination of the environment. That human tissue stores have changed so little during the past 20 years can be attributed mainly to strict enforcement of regulations minimizing chemical contamination of foodstuffs.

Proper statistical representation and comparison of data is as complex and important

Mean pesticide concentrations in five tissues in relation to mean total lipid contents.

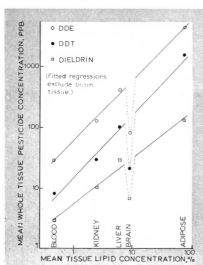

Table 6.—Means and Standard Deviations of Tissue Concentrations of Three Pesticides

	Blood	Adipose	Liver	Kidney	Brain
No. of specimens	37	70	70	66	52
Mean % lipid	0.44 ± 0.40	74.0 ± 17.0	5.8 ± 5.7	2.4 ± 2.3	8.9 ± 2.3
Mean pesticide concentration, ppm, in whole tissue					
DDT	0.0084 ± 0.0075	1.54 ± 1.65	0.101 ± 0.142	0.030 ± 0.056	0.022 ± 0.023
DDE	0.0291 ± 0.0354	4.58 ± 3.57	0.432 ± 0.518	0.133 ± 0.274	0.084 ± 0.066
Dieldrin	0.0034 ± 0.0030	0.14 ± 0.13	0.031 ± 0.064	0.011 ± 0.013	0.007 ± 0.007
Mean pesticide concentration, ppm, in extractable lipid					
DDT	2.22 ± 1.85	2.14 ± 2.30	1.91 ± 2.60	1.23 ± 1.39	0.26 ± 0.26
DDE	7.07 ± 6.47	6.07 ± 4.47	8.18 ± 8.00	5.27 ± 5.88	0.97 ± 0.73
Dieldrin	0.95 ± 0.82	0.20 ± 0.23	0.67 ± 1.31	0.52 ± 0.40	0.08 ± 0.14

as the development of chemical analytical methods. A marked tendency to skewness has been noted in tissue pesticide data, and the significance of this property has been examined.[10] A minority of exceptionally high values account for the large standard deviations characterizing our own data (Table 6). The geometric mean and the median are surely more representative of the central grouping in highly skewed distributions. When the arithmetic mean may be as much as five times the geometric mean, the selection of an appropriate statistic is, indeed, of more than trivial importance.[10] Unfortunately, statistical techniques applicable to medians and geometric means are not so widely used as those available for normally distributed data, and varying degrees of skewness in different distributions make the selection of appropriate statistical measures a matter of judgment.

Tissue pesticide levels in the chronically ill have attracted interest on the grounds that mobilization of fat stores may release a dangerous quantity of chemical into the tissues of the patient.

Estimation of the maximum available dosage from tissue stores should dispel fears of the first-mentioned type,[12] and extensive autopsy tissue studies have failed to correlate tissue pesticide stores with any of a number of disease processes or tissue injuries.[7] Our own data lend no support to the view that high tissue pesticide stores are associated with specific diseases or tissue injuries.

The lipotropic nature of chlorinated hydrocarbon pesticides is well demonstrated by the strong correlation between tissue lipid and pesticide content (Figure). However, in tissues whose lipid content is other than almost pure triglyceride (ie, nonadipose), an exact quantitative relationship between pesticide and lipid content is not predictable. The lipids in blood and parenchymal tissues exist in various complex relations to nonlipid constituents. For these lipoproteins, various pesticides probably have specific affinities. Finally, the quantities of lipid and pesticide extracted from a given tissue are certainly dependent on relative efficiencies of the several extraction systems available to the chemist.

These considerations bear directly on the issue of whether tissue pesticide measurements should be expressed in terms of "whole tissue weight", or "extractable lipid weight". The latter has been advocated on two grounds: (1) levels so expressed are more nearly comparable numerically, and (2) correlation of measurements in various tissue specimens should be improved by this correction.

Table 6 and data from other laboratories[3] demonstrate that the lipid correction serves the first purpose well, except in the case of brain. But this striking exception points up an important aspect of the lipid correction concept: factors more subtle than extractable lipid profoundly affect pesticide content, and cannot be ignored. While the lipid correction may serve a useful purpose in certain data comparisons, it may detract from other objectives.

The issue of correlation is actually much more complex. Measurement of total extractable lipid entails error, which, in some sets of tissue samples, may be as large as the real differences between specimens. In such cases, very little improvement in correlation of pesticide levels among tissues may be expected. In other data analyses, real differences may be large relative to error, and reduction of pesticide concentration to an extractable lipid basis effectively removes one source of variability. Table 5 tests empirically the value of correcting pesticide measurements to an "extractable lipid" basis, when correlating levels of DDT, DDE, and dieldrin in the five tissues represented in our autopsy sets. For DDT and DDE, the correction frequently enhances the degree of intertissue correlation, implying that tissue lipid measurement successfully takes into account a significant source of variation. In general, no such improvement is evident in the case of dieldrin. This might have been predicted from the weak correlation between tissue dieldrin and lipid levels (Table 4). It again suggests that nonlipid storage may be much more important in the case of dieldrin than it is for DDT and its metabolites.

Conclusions

Within the United States, no broad geographic differences in tissue pesticide storage levels are apparent. Adipose stores in Arizona residents have shown no increase in the past five years. Both observations conform with the view that tissue stores derive mainly from diet, and pesticide contamination of foodstuffs throughout the United States has been held below legally prescribed limits for the past 14 years.

No evidence has been developed that persons suffering chronic illness have exceptionally high tissue pesticide concentrations in the parenchymal organs, brain, fat, or blood.

There is good reason to believe that dieldrin is more extensively distributed to nonlipid cellular and blood components than are DDT and DDE.

Expression of tissue pesticide concentrations in terms of "total extractable lipid" may have statistical utility in particular data treatments. The calculation should not be regarded, however, as anything more than a crude reflection of biochemical reality. It ignores the heterogeneity of lipids in various tissues, the probable broad differences in distribution of pesticide chemicals to lipid and nonlipid cellular components, and it entails the introduction of an additional measurement error into the estimate of concentration.

This investigation was supported by contract No. 85-65-84 with the Division of Community Studies, Office of Product Safety, Bureau of Medicine, Food and Drug Administration, Consumer Protection and Environmental Health Service, Public Health Service, Atlanta.

Emmett H. Pachal and James A. Laubscher, MS, carried out the more than 300 tissue pesticide and lipid analyses. Ann Brooks performed the data analyses.

References

1. Mills PA: Collaborative study of certain chlorinated organic pesticides in dairy products. *J Assoc Official Agricultural Chemists* 44:171-177, 1961.

2. Mills PA, Onley JH, Gaither RA: Rapid method for chlorinated pesticide residues in nonfatty foods. *J Assoc Official Agricultural Chemists* 46:186-191, 1963.

3. Casarett LJ, Fryer GC, Yauger WL Jr, et al: Organochlorine pesticide residues in human tissue —Hawaii. *Arch Environ Health* 17:306-311, 1968.

4. Fiserova-Bergerova V, Radomski JL, Davies JE, et al: Levels of chlorinated hydrocarbon pesticides in human tissues. *Industr Med Surg* 36:65-70, 1967.

5. De Vlieger M, Robinson J, Baldwin MK, et al: The organochlorine insecticide content of human tissues. *Arch Environ Health* 17:759-767, 1968.

6. *Monitoring Food and People for Pesticide Content.* Scientific Aspects of Pest Control publication No. 1402, National Academy of Sciences-National Research Commission, 1966.

7. Hoffman WS, Adler H, Fishbein WI, et al: Relation of pesticide concentrations in fat to pathological changes in tissues. *Arch Environ Health* 15:758-765, 1967.

8. Dale WE, Quinby GE: Chlorinated insecticides in the body fat of people in the United States. *Science* 142:593-595, 1963.

9. Moss JA, Hathaway DE: Transport of organic compounds in the mammal: Partition of dieldrin and telodrin between the cellular components and soluble proteins of blood. *Biochem J* 91:384-393, 1964.

10. Robinson J: The burden of chlorinated hydrocarbon pesticides in man. *Canad Med Assoc J* 100:180-191, 1969.

11. Robinson J, McGill AEJ: Organochlorine insecticide residues in complete prepared meals in Great Britain during 1965. *Nature* 212:1037-1038, 1966.

12. Dale WE, Gaines TB, Hayes WJ Jr: Storage and excretion of DDT in starved rats. *Toxic Appl Pharmacol* 4:89-106, 1962.

Absorption, Storage, and Metabolic Conversion of Ingested DDT and DDT Metabolites in Man

Donald P. Morgan, MD, PhD, and Cifford C. Roan, PhD

AFTER a quarter century of use as a pesticide, DDT (1,1,1-trichloro-2,2-bis [*p*-chlorophenyl] ethane) usage is being sharply curtailed in the United States. Objectionable characteristics of the chemical are its stability and disposition to storage in animal tissues. First measured in the fat of a group of people in 1951,[1] concentrations have thus far shown no definite change over the years[2,3] As to the future, there is very little basis on which to predict changes in human stores of DDT and related materials. Recognizing the chemical's widespread environmental distribution, some human exposure will inevitably

---

The views expressed herein are those of the investigators and do not necessarily reflect the official viewpoint of the Food and Drug Administration.

continue for a long time to come. But even if there were no further addition of the chemical to human storage, it is not known at what rates DDT and related materials could be expected to disappear.

It is well established that mechanisms exist for chemical modification and excretion of DDT in animals[4-7] and in man.[8] Initial degradation of DDT may proceed either by dehydrochlorination, yielding the unsaturated DDE (1,1-dichloro-2,2-bis[*p*-chlorophenyl] ethylene) or by substitution of hydrogen for one chlorine atom, yielding the saturated DDD (1,1-dichloro-2,2-bis[*p*-chlorophenyl] ethane).[4] DDD readily degrades further through a series of intermediates to the excretable DDA (bis[*p*-chlorophenylacetic] acid), and is rarely found as a stored metabolite in the general population.

DDE, by contrast, apparently does not undergo further breakdown to DDA.[4] This stability accounts in large part, for the

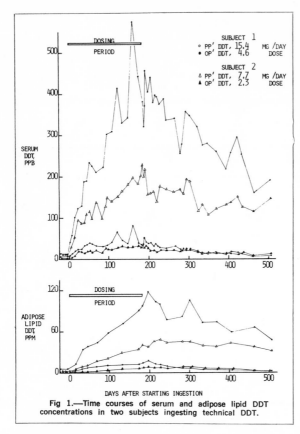

Fig 1.—Time courses of serum and adipose lipid DDT concentrations in two subjects ingesting technical DDT.

p,p'-DDE daily for 92 days, and a fourth 5 mg p,p'-DDD for 81 days.

Pesticide material for ingestion was first dissolved in vegetable oil. Appropriate quantities were then emulsified using water and acacia (Gum arabic) in proportions to yield the prescribed dose of chemical in either 2.5 or 5.0 ml of emulsion.[8] The dose of emulsion was taken daily, usually after meals. Blood samples were taken not less than eight hours after the last dose of pesticide.

A battery of hematologic and routine clinical biochemical tests were conducted before, during, and after the period of pesticide dosing to insure that the ingested material had no adverse effects. No abnormalities were detected in response to DDT, or to either of the metabolites.

Chemical Analysis

Serum samples were prepared for pesticide analysis as follows: Two milliliters of serum are added with stirring to 5 ml redistilled acetone in a beaker. Two milliliters of isooctane (2,2,4-trimethylpentane) are added, again with thorough mixing. Evaporation to near dryness is accomplished at 30 to 35 C, using dry air scrubbed through ethylene glycol. The material is then transferred quantitatively to a Mills clean-up column using n-pentane. (The column is prepared by 3 prerinses of a 25 ml mixture made up of the following: 15 parts ether[ethyl ether], 2 parts acetone, and 83 parts n-pentane).

Elution is achieved by 100 ml of the ether-acetone-pentane mixture described above; then the eluent is evaporated to near dryness, as before. Two milliliters of isooctane are now added, followed again by evaporation to near dryness. This final stage is then repeated to insure elimination of the lower-boiling solvents.

The material is now transferred to a grad-

higher human tissue store of DDE than of DDT in the general population.[2,3,9,10]

Some idea of the dynamics of adipose storage, metabolic conversion, and excretion of DDT in man can be obtained by following the time courses of DDT and metabolite levels in plasma and adipose fat before, during, and after a period of measured ingestion of DDT, or of a DDT metabolite. Such a study was undertaken in human volunteers.

Methods

Middle-aged adult men with no more than dietary and minimum household exposure to pesticides served as subjects. Two subjects ingested technical DDT (77%, p,p', 23% o,p' isomer): one 10 mg/day, the other 20 mg/day, for 183 days. Another subject consumed 5 mg uated centrifuge tube using ethyl ether rinses. One ml of isooctane is added, and evaporation is carried through to near dryness. Solution is brought to a convenient volume for gas-liquid chromatographic (GLC) analysis with pentane or isooctane.

Adipose samples were prepared for pesticide analysis as follows: Samples of 50 to 200 mg were removed by open biopsy from the upper and lateral buttock using 1% lidocaine (Xylocaine) anesthetic. Most of the early samples were stored by freezing prior to analysis; later samples were weighed and extracted immediately. Most of the specimens were analyzed in duplicate. Extraction was accomplished by homogenization in the presence of a 1:1:1 mixture of ethyl alcohol:hexane:ethyl ether. After washing with several volumes of water, the remaining extract was subjected to activated magnesium silicate (florisil) column cleanup, using as eluent a mixture of ethanol, ethyl ether, and pentane (0.3:14.7:85.0, v/v). Eluate was evaporated in dry air to a convenient volume for GLC analysis.

Adipose tissue concentrations of pesticide have been converted to *lipid* concentrations by dividing whole tissue values by the percent extractable lipid in each specimen.

GLC analysis for pesticides is carried out using columns six feet long with an inside diameter of one-quarter inch, containing 1.5% OV-17/2% QF-1 coating on 100 to 120 mesh chromatographic support, the latter having been acid and alcohol washed and hexamethyldisilazane treated.

Instrument operating parameters were as follows: inlet temperature, 230 C; detector temperature, 210 C; column temperature, 205 C; nitrogen flow, 20 ml/min; polarizing voltage, 10 to 40 volts direct current. Elec-

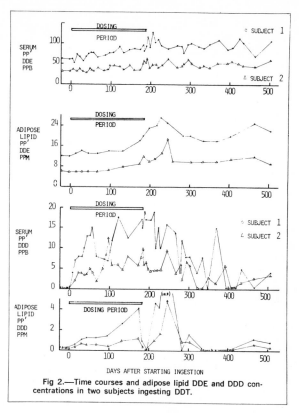

Fig 2.—Time courses and adipose lipid DDE and DDD concentrations in two subjects ingesting DDT.

tron capture detection has been employed throughout. Measurements are based on peak heights of chromatographic tracings.

Findings

Figure 1 shows changes in serum and adipose lipid concentrations of the two isomers of DDT in two subjects ingesting, respectively, 10- and 20-mg daily doses of technical DDT for six months.

Levels in serum and fat are seen to increase roughly in proportion to dosage. Increases in the *p,p'* isomer per unit dose far exceed elevations of the *o,p'* isomer. Blood and tissue levels of *o,p'*-DDT appear to approach a steady state in six months, while corresponding curves for the *p,p'* isomer bend only a little toward the time axis. The time courses for adipose lipid *p,p'*-DDT are

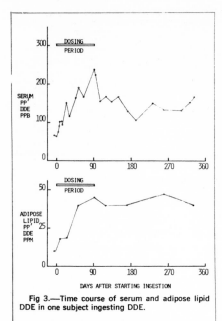

Fig 3.—Time course of serum and adipose lipid DDE in one subject ingesting DDE.

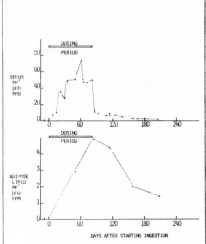

Fig 4.—Time course of serum and adipose lipid DDD in one subject ingesting DDD.

essentially linear over the six month interval.

It is noteworthy that adipose concentrations of p,p'-DDT in both subjects showed an early increase following termination of dosage, then declined in the course of the following nine months.

Figure 2 shows changes in serum and adipose levels of p,p'-DDE and p,p'-DDD corresponding to the same time course in the same subjects (ingesting technical DDT) described in Fig 1. The slow and modest changes in DDE during and after ingestion of DDT discount the view that very much of this metabolite found in man derives from absorbed DDT, as generally assumed or implied.[11,12] The increase, then fall, in DDE after the dosing period is not easily interpreted biologically, and may be an analytical artifact.

Analyses are adequate to establish that serum and adipose p,p'-DDD concentrations are elevated during DDT ingestion, roughly in proportion to dosage. Again, measurement error limits the interpretability of the plotted temporal fluctuations in concentration.

Figure 3 shows the rise in serum and adipose levels of DDE in a third subject in response to ingestion of 5 mg p,p'-DDE daily. Absorption and storage of this chemical are remarkably efficient: linearized rates of serum- and adipose tissue-concentration increase per unit dose are even higher in the case of DDE ingestion than for DDT ingestion.

Another comparison reveals the slowness of conversion of ingested DDT to DDE: DDE ingestion increased serum DDE levels 30 times as fast per unit dose as did DDT ingestion. Similarly, adipose lipid DDE increased 13 times as fast in response to DDE dosing as it did during DDT ingestion. In accord with this, Durham[7] found little or no conversion of ingested DDT to stored DDE in the monkey. This contrasts strikingly with the very prompt and efficient conversion found by Rothe in the rat.[5]

Similar study of the handling of orally administered DDD in a fourth subject demonstrates that DDD is absorbed and stored (Fig 4). The rapid elimination of tissue stores once dosing has stopped can be attributed mainly to the very efficient degradation of DDD to DDA, followed by excretion of the latter.[13]

58

Chemical Ingested	Subject No.	Duration of Ingestion, Days	Daily Dose, mg	Total Cumulative Dose, gm	Increment in Adipose Lipid Concentration, ppm	Estimated Adipose Lipid, kg*	Estimated Maximum Adipose Store, gm	% of Dose Stored
p,p'-DDT	1	183	15.4	2.82	113	17	1.92	68†
	2	183	7.7	1.41	42	18	0.76	54†
o,p'-DDT	1	183	4.6	0.84	16	17	0.27	32
	2	183	2.3	0.42	7	18	0.13	31
p,p'-DDE	3	92	5.0	0.46	32	13	0.42	91
p,p'-DDD	4	81	5.0	0.41	5	20	0.10	24

* Estimated by body specific gravity method of Behnke[14] using Rathbun-Pace equation.[15] Residual air was calculated as total lung capacity minus vital capacity; functional residual capacity was measured by nitrogen dilution.
† In both of these subjects, storage of p,p'-DDT as p,p'-DDE amounted to an additional 5% of ingested dose.

Table 2.—Estimates of adipose storage of DDT per Unit Oral Dose After Six Months of Dosing

No. of Subjects	Daily Dose Technical DDT, mg	Mean Adipose Total DDT at 6 Months (Uncorrected), ppm	Adipose Total DDT ppm/mg Dose	Reference
4	3.5	22*	6.3	Hayes[8]
7	35.0	139*	4.0	Hayes[8]
1	10.0	27	2.7	This paper
1	20.0	63	3.2	This paper

* Means calculated from points read from Fig. 2.[8]

We found no systematic change in serum or adipose DDD in the subject ingesting DDE, and no significant trend in DDE in the subject taking DDD. This tends to confirm the independence of these metabolic pathways.[1]

Comment

Storage Efficiency.—Table 1 indicates the maximum amounts of adipose-stored pesticide that could be accounted for during or after dosing in the four subjects. Admitting limited accuracy of body fat estimates made from specific gravity measurement,[11,15] it is none the less apparent that broad differences exist between storage efficiencies characterizing the isomers and metabolites. This property is evidently an important determinant of the relationship between dietary intake and adipose storage.[16]

Our whole-tissue adipose levels of DDT after six months of dosing are lower (corrected to unit dosage) than were reported by Hayes[8] in his subjects taking technical DDT (Table 2). The basis for the difference is not certain, but different methods of measurement have been shown to yield discrepancies of this magnitude and in this direction.[9]

Very possibly, earlier colorimetric methods have somewhat overestimated adipose storage of ingested DDT.

The consistent increase in adipose levels of DDT observed after dosing stopped requires explanation. Tissue penetration apparently proceeds nonuniformly, pesticide from other storage sites shifting into subcutaneous fat for a considerable time after oral intake has stopped. This phenomenon places limitations on the validity of pesticide measurements from a single adipose site, under conditions of increasing or decreasing storage levels.

Redistribution effects of this nature must also play a part in producing the very different time courses of serum and adipose saturations. First-order curves fitted to the serum p,p'-DDT curves during dosing suggest that 95% steady-state values would be reached in about 8½ months of dosing, while there is no indication of an approach to steady-state concentrations in subcutaneous fat tissue over the same period.

Pesticide Transport Into and Out of Storage.—Figures 5 and 6 trace continuously the relationships between serum and adipose concentrations of DDT isomers and metabolites before, during, and after the period of increased intake.

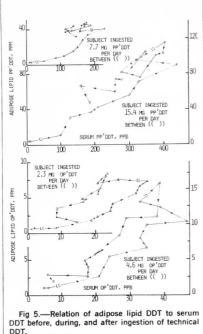

Fig 5.—Relation of adipose lipid DDT to serum DDT before, during, and after ingestion of technical DDT.

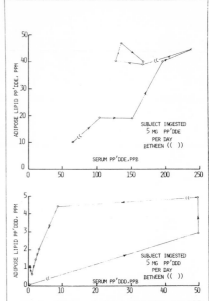

Fig 6.—Relation of adipose lipid DDE and DDD to the respective serum concentrations before, during, and after ingestion of these metabolites.

If rising and falling limbs of this curve were to superimpose, it could be inferred that equilibrium between serum and subcutaneous fat is attained about as fast as absorption and excretion change the serum concentration. Separation of the limbs of the "loop," on the other hand, implies that shifts into and out of subcutaneous fat do not keep pace with changes in serum levels.

For both DDT isomers, a modest gradient between storage elements is apparently necessary to effect transport of pesticides, but this is not so large as to impair seriously the use of serum levels to estimate concomitant adipose lipid concentrations. Approximate parallelism between on-dose and off-dose limbs of the loop suggests that storage and mobilization respond to similar dynamics.

The relationship for DDE is quite different. A prompt drop in serum level occurred after dosing terminated. This probably reflects final movement of serum DDE into adipose storage. Since this change, re-

moval from the blood has proceeded so slowly that the mobility of DDE out of adipose tissue cannot yet be evaluated.

Figure 3 has shown the prompt fall of serum DDD, once dosing with this chemical stopped. Figure 6 demonstrates mobilization from adipose tissue in response to the adipose-serum gradient.

Time Required for Excretion of Increments in DDT and DDT Metabolite Stores. —One of the principal objectives of this study has been to evaluate the time required for excretion of additional stores of pesticide generated during the period of measured ingestion.

Estimates made at only one year since dosing ended are obviously imperfect for the following reasons: (1) Steady states of blood and tissue concentrations did not exist when dosing ended, thus complicating interpretation of the early part of the curve. (2) Measurement error is large in relation to the systematic change in concentration over this

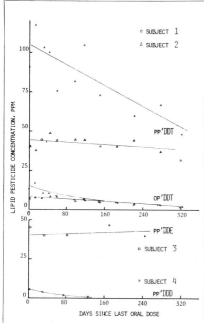

Fig 7.—Time courses of adipose lipid pesticide levels following final doses.

comparatively brief interval. Nor is it possible to generalize uncritically estimates made from so few subjects. One basic conclusion appears tenable even now, however: Excretion of an additional store of DDT or DDE, or both, proceeds much more slowly in man than it does in the monkey,[7] the dog,[17] and the rat.[18]

Data points available thus far on the subjects who consumed DDT, DDE, and DDD are presented in Fig 7. The points for DDT do not suggest departure from a linear decline in concentration with time. The rates of decline are strikingly different for the two subjects, no doubt due in part to individual metabolic differences. However, it is also well recognized that excretion of chlorinated hydrocarbons proceeds much more slowly from low storage levels than from high levels. In fact, not only does storage loss decline as the store is depleted, but time constants for logarithmic excretion curves fall progressively with storage level.[19] On this basis, it

can be anticipated that excretion of p,p'-DDT will decelerate from the mean rates observed in this first year since dosing ended.

Mean rates of p,p'-DDT loss from adipose storage, calculated from the regressions of Fig 7 have been 2.95 mg per day (0.15% of the additional store per day) in the high-dose subject and 0.35 mg per day (0.046% of the additional store per day) in the low-dose subject. These rates of loss substantially exceed what can be accounted for as urinary DDA during this period.[13] The alternative pathway of loss may be biliary, this mechanism having been well demonstrated in the rat.[20]

Removal of p,p'-DDE from fat is, if anything, slower than that of DDT. In eight months since dosing stopped, there has been no significant decline in adipose DDE in the subject who ingested this metabolite.

Lipid DDD, by contrast, declined promptly after dosage stopped. A fitted exponential curve suggests that even if excretion decelerates considerably, 95% of stored material should be lost within a year of the last dose.

Conclusions

Of the DDT-related materials studied, propensity for human adipose storage increases in this order p,p'-DDD $\leq$ o,p'-DDT $<$ p,p'-DDT $<$ p,p'-DDE. The chemicals stand in the same relationship with respect to stability of the adipose store, once dosage is terminated. Concentrations in serum and adipose samples from the general population are similarly related. Chemical stability in the body, efficiency of excretory mechanisms, and perhaps transport in and out of fat depots (especially in the case of DDE) are determinants of this relationship.

Serum and adipose concentrations of p,p'-DDE rise very slowly in response to DDT ingestion. This contrasts strikingly with the efficiency of absorption and storage of p,p'-DDE itself. Population levels of p,p'-DDE in blood and fat probably reflect primarily the person's absorption of preformed DDE itself, rather than DDT.

Mobilities of both isomers of DDT into and out of fat are adequate to yield a fair correspondence between serum and lipid concentrations even while stores are being built up or allowed to deplete. Less accurate correlation exists between serum and adipose levels in the cases of p,p'-DDE and p,p'-DDD, during storage and mobilization of these metabolites.

Loss of p,p'-DDT from adipose storage is much slower in man than in the monkey, dog, and rat. Many years will be required to remove 95% of increments in human stores built up by DDT and DDE ingestion. Stored p,p'-DDD is excreted much more rapidly than the same isomers of DDT and DDE.

The Arizona Community Studies Pesticide Project is supported by contract FDA-70-14 with the Division of Community Studies, Office of Pesticides and Product Safety, Bureau of Foods, Pesticides, and Product Safety, Food and Drug Administration, Public Health Service, US Department of Health, Education, and Welfare, Chamblee, Ga.

E. H. Paschal provided the pesticide measurements reported herein.

Louis J, Kettel, MD, chief of the Cardiopulmonary Laboratory, Veteran's Hospital, Tucson, Ariz, performed the lung volume measurements required in the calculation of body fat.

References

1. Laug EP, Kunze FM, Prickett CS: Occurrence of DDT in human fat and milk. *Arch Industr Hyg* **3:**245-246, 1951.

2. Quinby GE, Hayes WJ Jr, Armstrong JF, et al: DDT storage in the US population. *JAMA* **191:**109-113, 1965.

3. Hoffman WS, Adler H, Fishbein WI, et al: relation of pesticide concentrations in fat to pathological changes in tissues. *Arch Environ Health* **15:**758-765, 1967.

4. Peterson JE, Robison WH: Metabolic products of p,p'-DDT in the rat. *Toxic Appl Pharmacol* **6:**321-327, 1964.

5. Rothe CF, Mattson AM, Nueslein RM, et al: Metabolism of chlorophenothane (DDT). *Arch Industr Health* **16:**82-86, 1957.

6. Judah JD: Studies on the metabolism and mode of action of DDT. *Brit J Pharmacol* **4:**120-131, 1949.

7. Durham WF, Ortega P, Hayes WJ Jr: The effect of various dietary levels of DDT on liver function, cell morphology, and DDT, storage in the Rhesus monkey. *Arch Int Pharmacodyn* **141:**111-129, 1963.

8. Hayes WJ Jr, Durham WF, Cueto C Jr: The effect of known repeated oral doses of chlorophenothane (DDT) in man. *JAMA* **62:**890-897, 1956.

9. Dale WE, Quinby GE: Chlorinated insecticides in the body fat of people in the United States. *Science* **142:**593-595, 1963.

10. Morgan DP, Roan CC: Chlorinated hydrocarbon pesticide residue in human tissues. *Arch Environ Health* **20:**452-457, 1970.

11. Davies JE, Edmundson WF, Maceo A, et al: An epidemiologic application of the study of DDE levels in whole blood. *Amer J Public Health* **59:**435-441, 1969.

12. Edmundson WF, Davies JE, Nachman GA, et al: p,p'-DDT and p,p'-DDE in blood samples of occupationally exposed workers. *Public Health Rep* **84:**53-58, 1969.

13. Roan C, Morgan D, Paschal EH: Urinary excretion of DDA following ingestion of DDT and DDT metabolites in man. *Arch Environ Health* **22:**309-315, 1971.

14. Behnke AR Jr, Feen BG, Welham WC: The specific gravity of healthy men. *JAMA* **118:**495-501, 1942.

15. Rathbun EN, Pace N: Studies on body composition: The determination of body specific gravity. *J Biol Chem* **158:**667-676, 1945.

16. Durham WF, Armstrong JF, Quinby GE: DDT and DDE content of complete prepared meals. *Arch Environ Health* **11:**641-647, 1965.

17. Deichmann WB, Keplinger M, Dressler I, et al: Retention of dieldrin and DDT in the tissues of dogs fed aldrin and DDT individually and as a mixture. *Toxic Appl Pharmacol* **14:**205-213, 1969.

18. Datta PR, Nelson MJ: Enhanced metabolism of methyprylon, meprobamate, and chlordiazepoxide hydrochloride after chronic feeding of a low dietary level of DDT to male and female rats. *Toxic Appl Pharmacol* **13:**346-352, 1968.

19. Hayes WJ Jr: Review of the metabolism of chlorinated hydrocarbon insecticides, especially in mammals. *Ann Rev Pharmacol* **5:**27-52, 1965.

20. Jensen JA, Cueto C, Dale WE, et al: DDT metabolites in feces and bile of rats. *J Agric and Food Chem* **5:**919-925, 1957.

Urinary Excretion of DDA
Following Ingestion of DDT
and DDT Metabolites in Man

Clifford Roan, PhD; Donald Morgan, MD, PhD; and
Emmett H. Paschal

APPLE[1] presented a detailed review of metabolism and detoxification of DDT in mammals and new experimental results of studies with laboratory animals. Edmundson et al[2] have recently investigated DDA excretion in pesticide formulators. The most extensive earlier experiments employing human volunteers treated with controlled doses of DDT have been described by Hayes et al.[3] A 7½-year study with Rhesus monkeys on diets fortified with DDT has been reported by Durham et al.[4] These previous studies on the fate of DDT and related materials in man and other primates have not had the advantage of the sensitivity afforded by gas liquid chromatographic (GLC) techniques. Studies of urinary DDA excretion in human subjects ingesting measured doses of DDT or DDT metabolites have provided important clues to the metabolism of this chemical, and suggested methods by which future human exposure to DDT can be monitored.

Materials and Methods

Adult volunteers associated with the University of Arizona served as subjects. Details of treatment periods, compounds, and dosage appear in Table 1. The compounds were prepared as emulsions using the formulations described by Hayes et al.[3] The technical DDT was 77% *p,p'* isomer, 23% *o,p'* isomer. Only *pp'* isomers of DDE, DDD, and DDA were studied. The doses were selfadministered daily. One to two weeks prior to the start of dosing, then at intervals during and after dosing, samples of blood, urine, and adipose tissue were collected from each subject for biochemical and pesticide analysis. Urine samples were stabilized with formaldehyde solution, plus glacial acetic acid, or with concentrated hydrochloric acid. The exact time interval for each one day urine collection was recorded by the volunteer, from which an exact 24-hour excretion rate was calculated. Ten- to one hundred-ml aliquots (depending on anticipated concentration) were analyzed for DDA.

Table 1.—Human Volunteers Ingesting DDT and Related Chemicals

Subject*	Sex	Age (yr)	Height (in)	Weight (kg)	Chemical Ingested	Daily Dose	Period of Dosing (days)
1	M	31	68	77.1	p,p'-DDA	5 mg	21
2	M	31	75	104.3	p,p'-DDD	5 mg	81
3	M	41	71	73.0	p,p'-DDE	5 mg	92
4	F	52	64	59.8	Technical DDT	5 mg	52
5	M	48	69	71.2	Technical DDT	10 mg	183
6	M	44	70	70.3	Technical DDT	20 mg	183

* Subjects correspond with those in Fig 1 and 2.

DDA Analysis

The urine aliquot, with 3 ml concentrated H_2SO_4 in a 25-ml concentrator tube equipped with a concentrator column, was heated for one hour to 95 to 100 C on a water bath. After cooling to ambient temperatures the contents of the concentrator tube are extracted three times for one minute each with 5 ml of redistilled, peroxide free, ether (ethyl ether) using a high-speed mixer (Vortex). After separation, the upper layers in the concentrator tube are transferred to a culture tube (Pyrex) containing 1 ml of 2.5N NaOH. After capping the culture tube the combined ether extracts are mixed thoroughly with the NaOH layer, followed by evaporation of the ether layer from the culture tube using a stream of dry nitrogen. When evaporation is complete, 1 ml of dimethyl sulfate is added, the culture tube cap secured, and the mixture agitated in a mechanical shaker (30 oscillations per minute) overnight or until the brown color has migrated to the top layer. At this point 5 ml of a 1:1 mixture of ether and hexane are added followed by thorough mixing with a high-speed mixer. The contents of the culture tube are then transferred, using two rinses of 5 ml each of ether, then two rinses of 5 ml each of distilled water, to a separatory funnel containing 100 ml of distilled water. The contents of the separatory funnel are mixed thoroughly by five to ten seconds shaking, the phases allowed to separate and the water layer discarded. The ether layer is washed with water two additional times. The ether layer is transferred quantitatively, using ether rinses, to a 12-ml graduated stoppered centrifuge tube followed by evaporation to an appropriate volume for GLC analysis. The GLC analyses employ a microcoulometric detector. Operating specifications for the instrument are as follows: column (Pyrex), 6 feet × 0.25 inch (internal diameter); column packing, chromatographic support 100/120 mesh; column coating, 5% of a methyl silicone gum rubber (5% SE-30 after packing and curing, the column was treated with hexamethyldisilazane before use]). The inlet temperature was 235 C; column temperature, 210 C; and nitrogen flow, 125 ml/min. Quantitation was by peak height, the minimum detection limit, 1 ng/ml. The retention time for p,p'-DDA was 11 minutes.

Standards for o,p'-DDA determinations were not available during the technical DDT dosing period. Subsequently, the retention time for o,p'-DDA was determined as 9.4 minutes.

Results and Comment

The work of Hayes[3] indicated that DDA was the probable major metabolite of DDT to be found in urine. Laws et al[5] in investigating DDT workers found that of the probable metabolites of DDT, only DDA was related to occupational exposure.

Figure 1 shows the urinary excretion of p,p'-DDA in urine before, during, and after oral ingestion of this chemical. Excretion increased tenfold within six hours of taking the first dose. It remained high (average 42% of dose) as long as dosing continued. At the end of dosing, excretion dropped from 1.7 mg on the final day to 0.06 mg within 72 hours, after which daily excretions were all lower and within the range of predose values. These data suggest that for the general population and probably for occupationally exposed individuals, the ability of the body to excrete DDA does not limit removal of DDT from the body and that tissue storage of DDA is minimal.

Peterson and Robinson[6] has proposed that DDD derived by dechlorination of DDT undergoes further dechlorination to DDA by way of six intermediates. Our subject taking 5 mg daily doses of p,p'-DDD demonstrated rapid and sustained conversion of a large fraction (average 36%) of the ingested dose to DDA (Fig 1). There

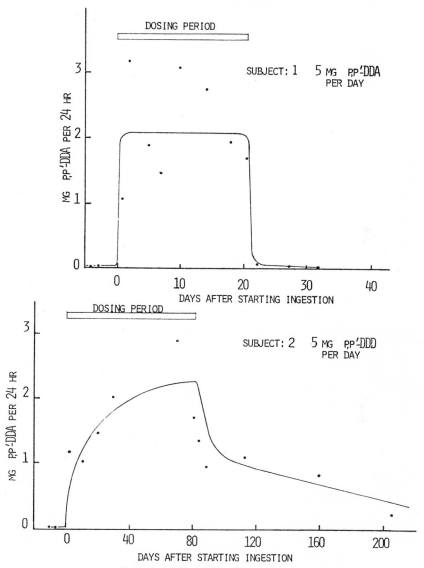

Fig 1.—Urinary excretion of **p,p'**-DDA before, during, and after oral treatment with **p,p'**-DDA or **p,p'**-DDD.

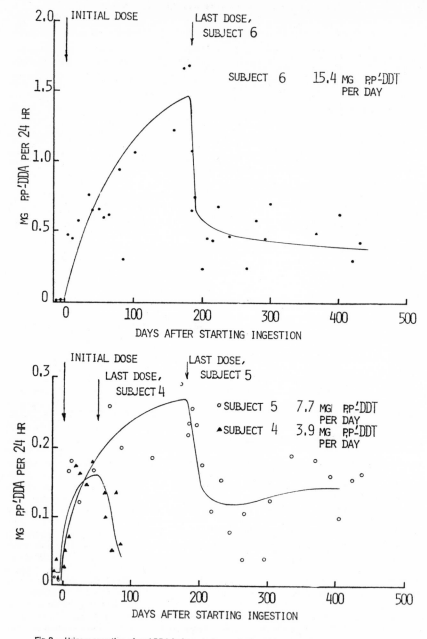

Fig 2.—Urinary excretion of p,p'-DDA before, during, and after oral treatment with technical DDT.

Table 2.—*Average Urinary Excretions of pp'DDA During Dosing Periods*

Ingested Chemical	Daily Dose As:		Mean Daily Excretion of DDA During Dosing, in Excess of Predose Rate*	
	Original Chemical	DDA Equivalent	Absolute (mg/day)	As % of Ingested DDA Equivalent
p,p'-DDA	5.0 mg	5.0 mg	2.1	42%
p,p'-DDD	5.0 mg	4.4 mg	1.6	36%
p,p'-DDE	5.0 mg	4.5 mg	0.0	0%
p,p'-DDT (in technical DDT)	3.9 mg	3.1 mg	0.11	4%
p,p'-DDT (in technical DDT)	7.7 mg	6.1 mg	0.20	3%
p,p'-DDT (in technical DDT)	·15.4 mg	12.2 mg	0.78	6%

* Total excretion rate estimated by planimetric integration of time-course excretion graphs. Basic dietary excretion of DDA, estimated from predose samples, subtracted from total excretion.

Table 3.—*Cumulative pp'DDA Excretion up to Nine Months From Start of Dosing**

Ingested Chemical	Total Ingested Dose (mg)		Total Urinary DDA Excretion in Excess of Predose Level	
	Original Chemical	DDA Equivalent	Absolute (mg)	As % of Ingested DDA Equivalent
p,p'-DDA	105	105	44	42.0%
p,p'-DDD	405	355	244	69.0%
p,p'-DDE	460	407	0	0.0%
p,p'-DDT (in technical DDT)	260	206	9†	4.4%
p,p'-DDT (in technical DDT)	1,410	1,115	50	4.5%
p,p'-DDT (in technical DDT)	2,820	2,230	219	10.0%

* Expressed as percent of total ingested DDA equivalent dose.
† Four months from start of dosing.

was a suggestion of improved DDA excretion with continued dosing. When dosing stopped, DDA excretion declined much less precipitously than in the subject who ingested DDA. Four months after dosing, DDA excretion is still five to six times the mean predose level for this subject. These changes correspond well with time course variations in his serum and adipose stores.[7] In fact, utilizing data from all subjects who took either DDT or DDD, a correlation coeficient of 0.80 can be calculated for the relationship between DDA excretion and serum DDD. The evidence that DDD is the immediate precursor of DDA appears to be strong.

White and Sweeney[3] in early investigations of DDT postulated on the basis of organic chemistry that DDE is an intermediate metabolite of DDT, ultimately degrading to DDA. Subsequently, Smith et al[9] observed that rats treated orally with DDE did not excrete DDA. Peterson and Robinson[6] likewise failed to detect DDA in urine of rats treated with DDE. Urine samples from the volunteer in our studies who ingested 5 mg p,p'-DDE daily for a period of 92 days have not contained any p,p'-DDA in excess of the levels found in pretreatment samples. This lack of significant urinary excretion of DDA derived from DDE partly explains the tendency of DDE in human adipose tissue to be remarkably stable.

Urines from subjects taking DDT and DDT metabolites were not routinely analyzed for DDT, DDE, or DDD, inasmuch as extraction procedures for isolation of DDA would exclude these materials. However, several 24-hour collections from subjects currently taking DDT and DDE were analyzed, using the classical Mills cleanup procedures for DDE in neutral urine extracts, and electron capture detection. These specimens yielded DDE concentrations ranging from "none detectable" to 0.9 parts per billion.

Four 100-ml urine samples collected during dosing from the volunteer taking DDE yielded 8 to 16 parts per billion of DDA, and 0.03 to 0.9 parts per billion of DDE. Of seven samples from the volunteers ingesting DDT, only three yielded detectable DDE. The mean DDE concentration in the samples was 0.27 parts per billion, giving a DDE to DDA ratio of 1:700.

Data from Laws et al[5] indicated a ratio of DDE to DDA in urine from occupationally exposed workers of 1:46. Neither the DDT nor the DDE excretion in these occupational groups appeared to be related to exposure, while DDA excretion appeared to be correlated. In view of our own limited data and the data from Laws,[5] we believe urinary excretion of DDA to be a better indicator of DDT and DDD exposure than urinary DDE or unmetabolized moieties. We could not demonstrate any relation between urinary DDE excretion and DDE intake.

Urinary p,p'-DDA excretion in volunteers ingesting technical DDT is shown in Fig 2. It is evident that excretory "ability" increases with continued dosing, and that upon termination, there is an abrupt decline. This feature of the time course is important, because the drop in DDA excretion was not associated with any corresponding decline in serum DDT levels.[7] The implication is strong that the higher levels of DDA excretion observed during dosing depended substantially on introduction of DDT directly into the gut. The demonstrated capacity of some gut bacteria to degrade DDT to DDD[10] reinforces the hypothesis that an important fraction of biodegradation by this pathway is actually enteric in location. Continuing biliary excretion of DDT[11] after dosing would deliver an adequate amount of DDT to the gut to account for the continuing "above normal" urinary DDA levels.

It is noteworthy that urinary DDA excretion does continue for months after dosing at levels several times greater than values reported for the general population. Durham et al,[12] using colorimetric measurement, found a range of 30 mg to 530μg per day in the general population. But Apple,[1] using a GLC method, calculated a mean of 13μg (range 3μg to 34μg) per day in general population samples, and 50μg (range 5μg to 207μg) per day in samples from 21 pest control operators. Fifteen predose 24-hour urine samples from six of our study subjects have yielded a mean population value of 16μg (range 4μg to 40μg) DDA per day. DDA excretions nearly always exceeding 50μg per day nine months after DDT dosing are distinguishable from these population levels.

What fraction of ingested DDT is excreted as urinary DDA? Duggan[13] presents data on intake of dietary residues by the general US population indicating average ingestion of 28μg and 14μg of DDT and DDD, respectively, or a total of 34μg of DDA equivalent material per day. On this basis, we would estimate that 47% of ingested precursor material is so excreted. During nine months following dosage of 7.7μg/day and 15.4μg/day of p,p'-DDT by the two subjects in this study, urinary p,p'-DDA excretion averaged 49% and 29%, respectively of estimated p,p'-DDT storage loss (corrected to pp'DDA).[7] In subjects brought to storage equilibrium by prolonged ingestion of 35μg per day of DDT, Hayes et al estimated that urinary DDA excretion accounted for 19% of the ingested dose.[3] Monkeys brought to very high steady state levels of DDT turnover (in relation to body weight) by prolonged dosage of technical DDT excreted only 3.4% of the ingested dose as urinary DDA.[4] Comparing these values, there is at least a strong suggestion that urinary DDA excretion becomes a relatively less important pathway of excretion as DDT intake reaches very high levels.

Because DDA derives only from DDT (and the less used chemical DDD) to the exclusion of DDE, measurement of DDA in the urine offers an obvious opportunity to assess prior human exposure to DDT, or DDD, or both. Urinary DDA excretion increases detectably within 24 hours of DDT ingestion. As a device for evaluating exposure this measurement may well prove simpler and more reliable than measurement of serum DDT concentrations.

Table 2 summarizes the efficiency of DDA synthesis and urinary excretion in subjects ingesting DDT or one of its metabolites. It is evident that the initial dechlorination of DDT is critical in determining how much and how rapidly DDT can be removed from the body by this route.

Table 3 shows cumulative urinary excretion of DDA up to nine months from the start of dosing, again expressed as a percent of ingested DDA equivalent. The higher recovery of DDD than of DDA itself implies more efficient intestinal absorption of the former. It must be remembered that most ingested DDT is still stored in the body at this time, and that the ultimate percentage disposition of this material as urinary DDA is not determinable as yet.

Our rates of urinary DDA excretion after dosing correspond satisfactorily with rates of mobilization of pesticide from adipose stores[7]: ie, loss of pesticide from storage was fastest in the subject who took DDD, slow in the DDT-dosed subject who has shown the lowest rate of DDA excretion, and not measurable (at eight months) in the DDE-dosed subject.

Conclusions

Ingested p,p'-DDA is promptly and efficiently excreted in the urine, undergoing virtually no tissue storage during ingestion.

Urinary excretion of p,p'-DDA increases during ingestion of p,p'-DDD to about the level generated by DDA itself. Additional DDD is absorbed and stored, yielding substrate for DDA synthesis after dosing is stopped.

Orally administered DDE produces no increase in urinary DDA excretion.

Urinary excretion of DDT as DDA appears to be totally dependent on the preferential reductive dechlorination of DDT to DDD (rather than DDE) and thence to DDA. DDA excretion rate increases as DDT ingestion continues. Urinary loss accounts for only a small fraction of ingested DDT up to nine months from the start of dosing. It accounts for more than half of ingested DDD in this time interval.

Urinary DDA excretion appears to account for a larger fraction of total DDT intake at low intake levels than at high levels.

When oral dosing of DDT stops, urinary DDA excretion drops off despite continuing high blood levels. An important role of gut flora in man's biodegradation of DDT is strongly suggested.

Measurement of urinary DDA excretion may offer a useful method of monitoring continued exposure of the general population to DDT, and of stratifying exposure levels among occupational users.

The Arizona Community Studies Pesticides Project is supported by Contract FDA-70-14 with the Division of Community Studies, Office of Pesticides and Product Safety, Bureau of Foods, Pesticides, and Product Safety, Food and Drug Administration, Public Health Service, US Department of Health, Education, and Welfare, Chamblee, Ga.

References

1. Apple E: *Metabolism and Detoxification of DDT in Mammals*, dissertation. University of Arizona Graduate College, Tucson, 1968.

2. Edmundson WF, Davies JE, Cranmer M, et al: Levels of DDT and DDE in blood and DDA in urine of pesticide formulators following a single intensive exposure. *Industr Med Surg* 38:55-60, 1969.

3. Hayes WJ, Durham WF, Cueto C: The effects of known repeated oral doses of chlorophenothane (DDT) in man. *JAMA* 162:890-897, 1956.

4. Durham WF, Ortega P, Hayes WJ: Effect of various dietary levels of DDT on liver function, cell morphology, and DDT storage in the Rhesus monkey. *Arch Int Pharmacodyn* 141:111-129, 1963.

5. Laws ER, Curley A, Biros FJ: Men with intensive occupational exposure to DDT. *Arch Environ Health* 15:765-775, 1967.

6. Peterson JE, Robinson WH: Metabolic products of pp'-DDT in the rat. *Toxic Appl Pharmacol* 6:321-327, 1964.

7. Morgan DP, Roan CC: Absorption, storage, and metabolic conversion of ingested DDT and DDT metabolites in man. *Arch Environ Health* 22:301-308, 1971.

8. White WC, Sweeney TR: Metabolism of DDT. I. A metabolite from rabbit urine, Di-(p chlorophenyl)-acetic acid: Its isolation, identification and synthesis. *Public Health Rep* 60:1-66, 1945.

9. Smith MI, Bauer H, Stohlman EF, et al: Pharmacologic action of certain analogues and derivatives of DDT. *J Pharmacol Exp Ther* 88:359-365, 1946.

10. Wedemeyer G: Dechlorination of 1,1,1-trichloro-2,2-bis (p-chlorophenyl) ethane by *Aerobacter aerogenes*: I. Metabolic products. *Appl Microbiol* 15:569-574, 1967.

11. Jensen JA, Cueto C, Dale WE, et al: DDT metabolites in feces and bile of rats. *Ag Food Chem* 5:919-925, 1957.

12. Durham WF, Armstrong JF, Quinby GE: DDA excretion levels. *Arch Environ Health* 11:76-79, 1965.

13. Duggan, RE: Residues in food and feed. *Pest Monitoring J* 2:2-46, 1968.

Physiological Effects of DDT on Man

Effect of intensive occupational exposure to DDT on phenylbutazone and cortisol metabolism in human subjects

Alan Poland, Donald Smith, R. Kuntzman, M. Jacobson, and A. H. Conney

Many drugs, carcinogens, insecticides, and steroids are hydroxylated by NADPH-dependent enzyme systems in liver microsomes and the terminal oxygenase for these enzymes is believed to be a carbon monoxide–binding cytochrome, termed cytochrome P-450.[15] Treatment of animals with

Supported in part by United States Public Health Service Grant HD-04313.

numerous drugs or chlorinated hydrocarbon insecticides stimulates drug and steroid hydroxylation in vitro, increases the amount of cytochrome P-450 in liver microsomes, causes proliferation of the smooth endoplasmic reticulum in liver, and decreases the action of drugs and steroids in vivo.[6, 9, 15, 20, 23, 26] Studies[2, 6, 7] in recent years have indicated that several drugs that stimulate drug and steroid metabolism in animals also exert these effects in man. DDT* and several closely related compounds are among the more potent stimulators of drug and steroid metabolism in animals.[9, 14, 16, 36] The minimum exposure to DDT required to stimulate the metabolism of pentobarbital and decrease its hypnotic action in rats results in levels of 10 to 15 μg of DDT per gram of fat,[13, 28] a concentration approaching that which commonly occurs in the human population.[11, 18, 24, 27] The present study was undertaken to determine whether people with intense and prolonged occupational exposure to DDT have altered drug and steroid metabolism. Since the disappearance of phenylbutazone from blood and the urinary excretion of 6β-hydroxycortisol have been utilized as tests for the induction of liver microsomal enzymes in animals and in man,[3, 6, 22] we have determined the serum half-life of phenylbutazone and the urinary excretion of 6β-hydroxycortisol in a control population and in people occupationally exposed to DDT in a DDT factory.

Methods

Selection of volunteers. The management of the Montrose Chemical Corporation of Torrance, California, supplied a list of employees working in their DDT plant for more than 5 years. All of these individuals had received moderate to intense occupational exposure to DDT, and all were in good health. Since DDT was the only product manufactured in the factory, there was little or no exposure to other chemicals in this plant. The starting materials for the synthesis of DDT were monochlorobenzene and trichloroacetaldehyde which were piped from tank cars to the reaction vessel, and the factory workers were not exposed to these substances. Employees were informed of the nature of the investi-

gation, and those who volunteered provided us with a medical history, donated blood for clinical laboratory determinations, and submitted to a physical examination. Special attention was given to drug idiosyncratic reactions, medication history, evidence or history of hepatic dysfunction, smoking and drinking habits, and contraindications to phenylbutazone administration. Those with normal liver function (as judged by normal serum concentrations of bilirubin, alkaline phosphatase, serum glutamic oxalacetic acid transaminase) and normal renal function (as judged by blood urea nitrogen) were admitted to the study. Eighteen male DDT factory workers with prolonged exposure to DDT (average duration of employment 14.4 ± 1.2 years*) were selected for the study.

A control population was selected from policemen and firemen in Brownsville, Texas, except for 2 persons who were officials from the United States Public Health Service in Atlanta, Georgia. Each volunteer in the control group was picked to match a DDT factory worker with respect to age, ethnic origin (Caucasian or Mexican-American), smoking habits, alcohol consumption, and medication. None of the factory workers or subjects in the control group received drugs known to affect drug metabolism. The control group was screened by medical history, physical examination, and laboratory tests and then submitted to the same studies as the DDT workers, except that the biopsy of adipose tissue was omitted. The average age of the DDT workers was 39.0 ± 1.5 years, and the average age of the control population was 40.2 ± 1.9 years. Each group contained 13 Mexican-American men and 5 Caucasian men.

Analytical methods. Adipose tissue was obtained by aspiration needle biopsy of the gluteal fat with a 17 gauge needle and placed in an airtight, tared vial.[24] Twelve

*The following abbreviations were used—DDT or p,p'-DDT: 1,1,1-trichloro-2,2 bis (p-chlorophenyl)ethane; o,p'-DDT: 1,1,1-trichloro-2-(o-chlorophenyl)-2-(p-chlorophenyl) ethane; p,p'-DDE: 1,1-dichloro-2,2 bis (p-chlorophenyl) ethylene; p,p'-DDD: 1,1-dichloro-2,2-bis (p-chlorophenyl) ethane.

*The mean ± standard error was used in reporting the data in this manuscript, and the Student's t test was used for the determination of significance.

fat samples, weighing 5 mg. or more (average weight, 11.9 mg.), were used for the analysis of DDT. Samples weighing less than 5 mg. were discarded. Adipose tissue and serum were frozen at –20° C. after collection and were thawed just prior to the analysis of DDT and related compounds by electron-capture gas-liquid chromatography as previously described.[10, 17] Serum was extracted with hexane, and fat samples were extracted with hexane-ether (3:1). The organic solvent was evaporated to dryness, and the residue was dissolved in a small volume of hexane before injection into the gas chromatography apparatus. Twenty-four hour urine samples were collected and chilled between collection periods. The 24 hour urine was then frozen at –20° C. until it was analyzed for 6β-hydroxycortisol by modifications[22] of earlier methods.[12, 37] After the samples were obtained, the volunteers were given a single 400 mg. oral dose of phenylbutazone, and serum was obtained at 24, 48, 72, 96, and 120 hours after drug administration. The serum samples were frozen at –20° C. until they were analyzed for phenylbutazone as described earlier.[4] Each analysis was done in duplicate, and the

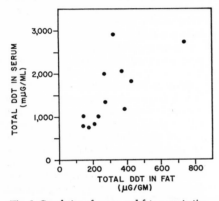

Fig. 1. Correlation of serum and fat concentration of total DDT-related compounds (*see* Table I) in 12 DDT factory workers. The correlation coefficient was r = 0.72 and p < 0.01.

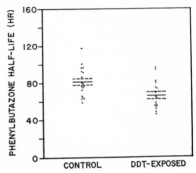

Fig. 2. The serum half-life of phenylbutazone in a control population and in DDT factory workers. The individual values and mean ± standard error are shown.

Table I. *Serum and fat concentrations of DDT and related compounds in a control population*

| Population studied | Serum concentration* of DDT and related compounds (mμg/ml. or p.p.b.) | | | | |
	Sum of DDT and related compounds (expressed as p, p'-DDT)	p,p'-DDT	p,p'-DDE	o,p'-DDT	p,p'-DDD
Control group	51 ± 5 (range, 24-102)	12 ± 2	35 ± 3	< 2†	< 2‡
DDT factory workers	1,359 ± 162 (range, 579-2,914)	573 ± 60	506 ± 88	80 ± 14	97 ± 14

*The serum concentrations represent the mean and standard error from 18 samples in each group, and the adipose tissue
†In 16 samples, the concentration of o,p'-DDT was undetectable, < 2 mμg per milliliter. In the samples in which it was
‡In 15 samples, the concentration of p,p'-DDD was < 2 mμg per milliliter. In the other 3, the concentrations were 4.5,

phenylbutazone half-life was determined by the regression line and fitted by the method of least squares, with 4 to 5 points for each half-life.

Results

Concentration of DDT in control population and DDT factory workers. The average serum concentrations of p,p'-DDT, p,p'-DDE, o,p'-DDT, p,p'-DDD, and the sum of all DDT-related compounds expressed as p,p'-DDT were substantially higher in the DDT factory workers than in the control group (Table I). The average serum concentration of all DDT-related compounds expressed as p,p'-DDT was 51 ± 5 mµg per milliliter (range 24 to 102 mµg per milliliter) in the control group and 1,359 ± 162 mµg per milliliter (range 579 to 2,914 mµg per milliliter) in the factory workers. The analysis of 12 fat samples from DDT workers revealed an average fat concentration of 307 ± 48 µg per gram for the sum of all DDT-related compounds (Table I). This is 20 to 30 times the concentration reported for the general population in the United States.[11,][18, 24, 27] Similarly, much higher concentrations of p,p'-DDT, p,p'-DDE, o,p'-DDT, and p,p'-DDD were found in the fat of the DDT factory workers than were reported previously for the general population.[27] There was a significant correlation (r = 0.72; $p < 0.01$) between the serum and fat concentration of total DDT-related compounds in the DDT factory workers (Fig. 1).

Effect of DDT exposure on phenylbutazone half-life and urinary excretion of 6β-hydroxycortisol. The data in Table II and Figs. 2 and 3 suggest that heavy exposure to DDT significantly ($p < 0.01$) shortens the phenylbutazone half-life and increases the urinary excretion of 6β-hydroxycortisol. The serum half-life of phenylbutazone was 81.0 ± 3.7 hours in the control group and 65.5 ± 3.5 hours in the DDT workers. The average urinary excretion of 6β-hydroxycortisol during a 24 hour interval was

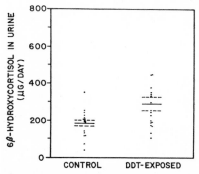

Fig. 3. The 24 hour urinary excretion of 6β-hydroxycortisol in a control population and in DDT factory workers. The individual values and mean ± standard error are shown.

and in DDT factory workers

Adipose tissue concentration° of DDT and related compounds (µg/Gm. or p.p.m.)				
Sum of DDT and related compounds (expressed as p,p'-DDT)	p,p'-DDT	p,p'-DDE	o,p'-DDT	p,p'-DDD
Not measured				
307 ± 48 (range, 141-739)	148 ± 26	91 ± 17	24 ± 3	34 ± 4

concentrations are the mean and standard error from 12 DDT workers.
detected, the values were 2.9 and 2.4 mµg per milliliter.
5.7, and 2.5 mµg per milliliter.

Table II. *The serum half-life of phenylbutazone and the 24 hour urinary excretion of 6β-hydroxycortisol in a control population and in DDT factory workers*

Population studied	Subjects	6β-Hydroxycortisol excretion (μg/24 hr.)	Phenylbutazone half-life (hr.)	Total DDT in serum[a] (mμg/ml.)
Control group	S. C.	41	91.9	26
	P. A.	76	96.3	35
	G. O.	118	79.7	41
	N. O.	118	92.4	102
	E. V.	133	65.7	75
	C. A.	144	96.3	58
	Z. A.	191	76.2	38
	S. A.	193	76.2	52
	M. U.	200	63.2	88
	T. R.	200	59.2	40
	G. U.	205	72.2	79
	D. L.	208	64.2	30
	C. R.	208	117.5	56
	L. U.	213	63.6	24
	C. H.	229	80.6	35
	Z. M.	245	75.3	42
	L. O.	255	86.6	61
	B. E.	355	100.4	39
DDT factory workers	D. I.	106	48.4	751
	H. O.	134	69.2	781
	C. L.	171	62.9	579
	P. A.	173	55.7	1,001
	G. A.	190	95.7	1,172
	A. Y.	192	45.9	826
	G. R.	198	73.3	920
	P. L.	231	63.9	816
	R. O.	248	79.7	2,049
	V. I.	254	54.6	2,725
	Z. A.	300	61.7	2,914
	R. A.	335	58.0	1,013
	P. R.	345	51.4	835
	B. U.	351	74.3	1,986
	T. I.	380	81.3	1,382
	A. A.	447	53.8	1,809
	V. E.	449	94.3	1,335
	Z. M.	741	54.9	1,565
Average ± S.E.				
Control group		185 ± 17	81.0 ± 3.7	51 ± 5
DDT factory workers		291 ± 36	65.5 ± 3.5	1,359 ± 162

[a] Sum of DDT and related compounds expressed as p,p'-DDT.

185 ± 17 μg in the control group and 291 ± 36 μg in the DDT factory workers. A significant correlation did not occur between the serum concentration of total DDT-related compounds and the urinary excretion of 6β-hydroxycortisol in the 18 DDT factory workers ($r = 0.32$; $p >$ 0.05), or in the 18 control subjects ($r = -0.07$; $p > 0.05$), (Table II). Although the serum half-life of phenylbutazone was significantly shorter in DDT factory workers than in the control population, the serum half-life of phenylbutazone did not correlate with the total concentration of DDT and

related compounds in the serum of either the DDT factory workers ($r = 0.07$; $p > 0.05$) or the control group ($r = 0.02$; $p > 0.05$) (Table II). Similarly, the fat concentration of total DDT-related compounds in DDT factory workers did not correlate with the phenylbutazone half-life or the urinary excretion of 6β-hydroxycortisol (Fig. 1, Table II). In addition, the biological half-life of phenylbutazone in different individuals did not correlate with the urinary excretion of 6β-hydroxycortisol in the control population ($r = -0.07$; $p > 0.05$) or the DDT factory workers ($r = 0.02$; $p > 0.05$) (Table II).

Discussion

The studies described here indicate that people with prolonged occupational exposure to large amounts of DDT have a decreased serum half-life of phenylbutazone and an increased urinary excretion of 6β-hydroxycortisol. Similar studies by Kolmodin and associates[19] showed that people occupationally exposed to a mixture of halogenated hydrocarbon insecticides (primarily lindane, DDT, and chlordane) have enhanced metabolism of antipyrine. The effects observed in our study and in the study by Kolmodin and co-workers were probably caused by the induction of liver microsomal enzymes that metabolize phenylbutazone, antipyrine, and cortisol, since long-term treatment of animals with DDT or other halogenated hydrocarbon insecticides stimulates the hydroxylation of several drugs and steroid hormones by enzymes in liver microsomes.[9, 14, 16, 23, 36] The stimulatory effect of DDT on phenylbutazone metabolism and cortisol hydroxylation reported here is similar to the effect of other liver microsomal enzyme inducers on the metabolism of phenylbutazone and cortisol in man. Treatment of people with barbiturates or certain other drugs lowered the average phenylbutazone half-life from 78 hours to 57 hours in a normal population and from 100 hours to 54 hours in people with liver disease.[25] Treatment of people with phenobarbital,[5] N-phenylbarbi-

tal,[22] diphenylhydantoin,[37] phenylbutazone,[21] or o,p′-DDD[1, 29, 30] increased the urinary excretion of 6β-hydroxycortisol, but these increases were greater than those found in DDT factory workers. Several of the drugs that enhanced the urinary excretion of 6β-hydroxycortisol in man have been shown to stimulate the activity of an enzyme system in guinea pig liver microsomes that hydroxylates cortisol in the 6β-position.[8, 21, 22] These observations suggested that measurement of the urinary excretion of 6β-hydroxycortisol may be a useful test for the induction of liver microsomal enzymes in man.

The serum concentration and fat stores of DDT-related substances in DDT factory workers were 20 to 30 times those in the control population, but the urinary excretion of 6β-hydroxycortisol was increased only 57 per cent, and the serum phenylbutazone half-life was reduced by only 19 per cent. These findings suggest that intense and prolonged exposure to DDT stimulates hepatic drug and steroid metabolism in man, but the effect on phenylbutazone metabolism and cortisol hydroxylation is a modest one. Although we do not know whether DDT storage in the general population is sufficient to stimulate phenylbutazone metabolism or the urinary excretion of 6β-hydroxycortisol, the extent of the possible induction appears to be small. The data in Table II indicate considerable variability in the serum half-life of phenylbutazone and the urinary excretion of 6β-hydroxycortisol in different people. The variability in the concentration of DDT in the tissues of different individuals is not the major reason for individual differences in the rates of hydroxylation of phenylbutazone or cortisol in man, since the serum concentration of DDT-related substances in either the control population or the DDT factory workers did not correlate with the serum half-life of phenylbutazone or the urinary excretion of 6β-hydroxycortisol. These observations suggest that variability in the serum half-life of phenylbutazone and

77

variability in the urinary excretion of 6β-hydroxycortisol in the normal population or in the DDT factory workers results primarily from genetic factors or environmental factors unrelated to DDT exposure.

The lack of correlation between serum DDT levels and the serum half-life of phenylbutazone or the urinary excretion of 6β-hydroxycortisol may be explained by genetic differences in susceptibility to enzyme induction in different individuals. Recent studies by Vesell and Page[34] have pointed out the importance of genetic factors in determining the stimulatory effect of phenobarbital on antipyrine metabolism in man. Marked individual differences in the induction of placental benzpyrene hydroxylase by cigarette smoke has also been observed. Although little or no benzpyrene hydroxylase activity is present in the placentas from nonsmokers, the placentas from cigarette smokers have appreciable benzpyrene hydroxylase activity and there was a 25-fold difference in benzpyrene hydroxylase activity in the placentas from different women who smoked the same number of cigarettes.[35]

The importance of genetic factors in explaining individual differences in the rates of metabolism of several drugs that are metabolized by liver microsomal enzymes was recently pointed out by Vesell and associates.[31, 32, 33] These investigators studied individual variability in the metabolism of phenylbutazone, antipyrine, and bishydroxycoumarin in pairs of identical and fraternal twins and concluded that the rates of metabolism of these drugs in healthy people not treated with enzyme inducers or enzyme inhibitors was primarily under genetic rather than environmental control. Large differences in the half-lifes of the drugs among unrelated individuals disappeared almost completely in identical twins but persisted to some extent in most sets of fraternal twins. The relative roles of genetic and environmental factors in controlling the rates of metabolism of various commonly used drugs is an important area of pharmacology that is under investigation in many laboratories.

Although the serum half-life of phenylbutazone was significantly shorter and the urinary excretion of 6β-hydroxycortisol was significantly elevated in the DDT workers compared to the control group, within each group of 18 people there was no correlation between the rate of phenylbutazone disappearance and the excretion of 6β-hydroxycortisol in the urine. Phenylbutazone metabolism and 6β-hydroxycortisol formation involve hydroxylation by hepatic microsomal enzymes in animals, but the data obtained in our study suggest that phenylbutazone and cortisol are hydroxylated by different systems that are under separate regulatory control. Recent studies[32] indicate that the rates of metabolism of phenylbutazone and antipyrine are not correlated in different people, but the rate of phenylbutazone metabolism is correlated with the rate of bishydroxycoumarin metabolism,[33] suggesting that the metabolism of the latter two drugs is under similar regulatory control. Additional studies are needed to determine the number of drug-metabolizing systems that exist in man and the various factors that influence their activities.

We want to thank the management of Montrose Chemical Corporation, Torrance, Calif., for making this study possible. We also want to thank Mrs. Patricia Hickman and Miss Janet King for their expert technical assistance, and Mr. Ronald Gauch for the statistical analysis of the data.

References

1. Bledsoe, T., Island, D. P., Ney, R. L., and Liddle, G. W.: An effect of o,p'-DDD on the extra-adrenal metabolism of cortisol in man, J. Clin. Endocr. 24:1303-1311, 1964.
2. Burns, J. J.: Implications of enzyme induction for drug therapy, Amer. J. Med. 37:327-331, 1964.
3. Burns, J. J., Cucinell, S. A., Koster, R., and Conney, A. H.: Application of drug metabolism to drug toxicity studies, Ann. N. Y. Acad. Sci. 123:273-286, 1965.
4. Burns, J. J., Rose, R., Chenkin, T., Goldman, A., Schubert, A., and Brodie, B. B.: The physiological disposition of phenylbutazone (Butazolidin) in man and a method for its estima-

tion in biological material, J. Pharmacol. Exp. Ther. 109:346-357, 1953.

5. Burstein, S., and Klaiber, E. L.: Phenobarbital-induced increase in 6β-hydroxycortisol excretion: Clue to its significance in human urine, J. Clin. Endocr. 25:293-296, 1965.

6. Conney, A. H.: Pharmacological implications of microsomal enzyme induction, Pharmacol. Rev. 19:317-366, 1967.

7. Conney, A. H.: Drug metabolism and therapeutics. Seminars in Medicine of the Beth Israel Hospital, Boston, New Eng. J. Med. 280: 653-660, 1969.

8. Conney, A. H., Jacobson, M., Schneidman, K., and Kuntzman, R.: Induction of liver microsomal cortisol 6β-hydroxylase by diphenylhydantoin or phenobarbital: An explanation for the increased excretion of 6β-hydroxycortisol in humans treated with these drugs, Life Sci. 4:1091-1098, 1965.

9. Conney, A. H., Welch, R. M., Kuntzman, R., and Burns, J. J.: Effects of pesticides of drug and steroid metabolism, CLIN. PHARMACOL. THER. 8:2-10, 1967.

10. Dale, W., Curley, A., and Hayes, W.: Determination of chlorinated insecticides in human blood, Industr. Med. Surg. 36:275-280, 1967.

11. Durham, W. F.: Pesticide exposure levels in man and animals, Arch. Environ. Health 10: 842-846, 1965.

12. Frantz, A., Katz, F., and Jailer, J.: 6β-Hydroxycortisol and other polar corticosteroids: Measurement and significance in human urine, J. Clin. Endocr. 21:1290-1303, 1961.

13. Gerboth, G., and Schwabe, U.: Einflusz von gewebsgespeichertem DDT auf die Wirkung von Pharmaka, Naunyn-Schmildeberg Arch. Exp. Path. 246:469-483, 1964.

14. Gillett, J.: "No effect" level of DDT in induction of microsomal epoxidation, J. Agric. Food Chem. 16:295-297, 1968.

15. Gillette, J. R., Conney, A. H., Cosmides, G. J., Estabrook, R. W., Fouts, J. R., and Mannering, G. J., editors: Microsomes and drug oxidations, New York, 1969, Academic Press, Inc.

16. Hart, L., and Fouts, J.: Further studies on the stimulation of hepatic microsomal drug-metabolizing enzymes by DDT and its analogs, Naunyn-Schmiedeberg Arch. Exp. Path. 249:486-500, 1965.

17. Hayes, W., Dale, W., and Burse, V.: Chlorinated hydrocarbon pesticides in the fat of people in New Orleans, Life Sci. 4:1611-1615, 1965.

18. Hayes, W., Quinby, G., Walker, K., Elliot, J., and Upholt, W.: Storage of DDT and DDE in people with different degrees of exposure to DDT, Arch. Industr. Health 18:398-406, 1958.

19. Kolmodin, B., Azarnoff, D. L., and Sjöqvist, F.: Effect of environmental factors on drug metabolism: Decreased plasma half-life of antipyrine in workers exposed to chlorinated hydrocarbon insecticides, CLIN. PHARMACOL. THER. 10:638-642, 1969.

20. Kuntzman, R.: Drugs and enzyme induction, Ann. Rev. Pharmacol. 9:21-36, 1969.

21. Kuntzman, R., Jacobson, M., and Conney, A. H.: Effect of phenylbutazone on cortisol metabolism in man, Pharmacologist 8:195, 1966.

22. Kuntzman, R., Jacobson, M., Levin, W., and Conney, A. H.: Stimulatory effect of N-phenylbarbital (phetharbital) on cortisol hydroxylation in man, Biochem. Pharmacol. 17:565-571, 1968.

23. Kupfer, D.: Effects of some pesticides and related compounds on steroid function and metabolism, Residue Rev. 19:11-30, 1967.

24. Laws, E., Curley, A., and Biros, F.: Men with intensive occupational exposure to DDT, Arch. Environ. Health 15:766-775, 1967.

25. Levi, A. J., Sherlock, S., and Walker, D.: Phenylbutazone and isoniazid metabolism in patients with liver disease in relation to previous drug therapy, Lancet 1:1275-1279, 1968.

26. Mannering, G. J.: Significance of stimulation and inhibition of drug metabolism in pharmacological testing, in Selected pharmacological testing methods, vol. 3, New York, 1968, Marcel Dekker, Inc., p. 51.

27. Quinby, G., Hayes, W., Armstrong, J., and Durham, W.: DDT storage in the U. S. population, J. A. M. A. 191:175-179, 1965.

28. Schwabe, U.: DDT-Speicherung bei der Haltung von Versuchstieren als mogliche Fehlerguelle bei Arzneimittelprufungen, Arzneimittelforschung 14:1265-1266, 1964.

29. Southren, A. L., Tochimoto, S., Isurugi, K., Gordon, G. G., Krikun, E., and Stypulkowski, W.: The effect of 2,2-bis(2-chlorophenyl-4-chlorophenyl)-1,1-dichloroethane (o,p'-DDD) on the metabolism of infused cortisol-7-³H, Steroids 7:11-29, 1966.

30. Southren, A. L., Tochimoto, S., Stron, L., Ratuschni, A., Ross, H., and Gordon, G.: Remission in Cushing's syndrome with o,p'-DDD, J. Clin. Endocr. 26:268-278, 1966.

31. Vesell, E. S., and Page, J. G.: Genetic control of drug levels in man: Phenylbutazone, Science 159:1479-1480, 1968.

32. Vesell, E. S., and Page, J. G.: Genetic control of drug levels in man: Antipyrine, Science 161:72-73, 1968.

33. Vesell, E. S., and Page, J. G.: Genetic control of Dicumarol levels in man, J. Clin. Invest. 47:2657-2663, 1968.

34. Vesell, E. S., and Page, J. G.: Genetic control of the phenobarbital-induced shortening of

plasma antipyrine half lives in man, J. Clin. Invest. 48:2202-2209, 1969.

35. Welch, R. M., Harrison, Y. E., Gommi, B. W., Poppers, P. J., Finster, M., and Conney, A. H.: Stimulatory effect of cigarette smoking on the hydroxylation of 3,4-benzpyrene and the N-demethylation of 3-methyl-4-monomethylaminoazobenzene by enzymes in human placenta, CLIN. PHARMACOL. THER. 10:100-109, 1969.

36. Welch, R. M., Levin, W., and Conney, A. H.: Insecticide inhibition and stimulation of steroid hydroxylase in rat liver, J. Pharmacol. Exp. Ther. 155:167-173, 1967.

37. Werk, E. E., Jr., MacGee, J., and Sholiton, L. J.: Effect of diphenylhydantoin on cortisol metabolism in man, J. Clin. Invest. 43:1824-1835, 1964.

Evidence of Safety of Long-Term, High, Oral Doses of DDT for Man

Wayland J. Hayes, Jr., MD, PhD; William E. Dale; and Carl I. Pirkle, MD

Fifty-ONE MEN participated in an earlier investigation of the effect of known, repeated, oral doses of DDT.[1] Three of these men completed one year of dosage at 3.5 mg per man per day and seven men completed one year at 35 mg per man per day. The larger dosage was about 200 times the daily rate at which the average person received DDT from his diet at that time.[2] In this limited study, no evidence of injury related to DDT was reported by the men or was found by careful medical examination. The storage of DDT was proportional to dosage. DDE, a metabolite of DDT, constituted about 20% of the total storage in contrast to about 60% in the general population at that time. In the course of the study, a method was found[3] for the determination of DDA, a urinary metabolite of DDT, and preliminary measurements were made.

A second investigation was carried out with volunteers in order to: (a) reevaluate the earlier findings, (b) study the storage of DDT over a period considered long enough to permit establishment of a steady state, (c) study storage loss of DDT for the first time in man, and (d) measure the excretion of urinary DDA in detail.

The last samples collected in the second study were taken on Feb 6, 1961, and analyzed soon afterward. Most of the results were presented in a working paper prepared for the World Health Organization.[4] Publication was delayed in the hope of following at least some of the subjects for a longer period. The latest exchange of correspondence with one of them was in February 1969. However, no medically significant information could be obtained after the last regular examination. The results of the study as they relate to storage and excretion have been referred to in some detail, not only in the working paper but also in several published papers.[5-7] A full account of the work is considered desirable in spite of the delay and the fact that some toxicologists are already familiar with the results.

Methods

Permission was obtained from the Bureau of Prisons, US Department of Justice, to carry out a new study using volunteers in a penitentiary. An announcement of the study was posted on bulletin boards, and those inmates who were interested attended a meeting at which the plans were described in detail and questions were answered. The men who tentatively volunteered were examined and were accepted if they were qualified physically and in other ways. Participation was entirely voluntary. Each man was free to leave the study at any time. A further evidence that the work was voluntary is the fact that several men continued to submit fat samples on schedule after they were released from prison and had moved to other parts of the country. In these instances, the volunteer went to the physician of his choice who took and shipped the biopsy specimen according to our directions.

Forty-three men volunteered for the new study. The men were assigned case numbers after they had been placed in four groups by dealing cards bearing their names and arranged according to their ages. The design of the investigation is shown in Table 1. During the period of February 13 to 17, 1956, the initial examinations were made and biopsy specimens were taken. Administration of emulsion (placebo) containing no DDT was started on Feb 20, 1956. Without announcement, DDT was first included in the emulsion of appropriate groups on March 1, 1956.

Seven of the men were removed from the project at their own request during the third through the eighth month of dosing. Two of these men had received a daily dose of 35 mg of DDT; one man had received 3.5 mg; and four men had received no DDT except the trace present in the ordinary diet. These men left the study for administrative reasons (chiefly because the project interfered with their other duties) and not because of any injury by DDT. Two men became sick and their dosing was stopped, although it was not considered that their illness was related to DDT intake. Ten other men were released from custody or transferred before the project was finished. There remained 24 men who completed the project; the results from these men are used as the statistical basis of the major portion of this report dealing with storage and excretion. With two exceptions (both in the control group), each of the 24 volunteers just mentioned contributed eight biopsy specimens of subcutaneous fat from the abdomen and a little over 90 samples of urine during the four-year study. Two men contributed daily samples of urine for a year.

Although the paper is based primarily on 24 men studied for four years, all significant findings are reported for the men who, for any reason, failed to complete the entire project. Information is given also on the results of analysis of an additional biopsy and about 11 additional samples of urine from 16 of the 24 men who could be studied for one additional year after the project was formally terminated.

Group 1 was a control and received no DDT except the trace present in the regular diet. This dietary level was not measured during the study but was estimated at 0.18 mg per man per day on the basis of measurements by Walker et al.[2] Men in groups 2 and 3 each received daily doses of 3.5 and 35 mg, respectively, of technical DDT containing approximately 85% of the p,p'-isomer. Each man in group 4 received 35 mg daily of recrystallized p,p'-DDT. Each man received daily about 2 oz of milk to which had been added peanut oil emulsion containing the appropriate amount of DDT. This was the dosage technique used in the latter part of the earlier study. Unfortunately it was not possible to see to it that every man got his dose every day as was possible in the earlier study. Every volunteer missed some doses. There was also objective evidence that the doses were occasionally mixed so that one man got another's dose.

Oher procedures directly relevant to DDT and most of the clinical laboratory procedures were the same as those used in the earlier study.[1] A few additional tests were done, some at only one or two examinations during the course of the study.

Albert J. Schneider, MD, determined the serum glutamic oxaloacetic transaminase (SGOT) activity. Schneider and Willis[8] described the method used.

Robert F. Witter, PhD, was responsible for measuring the carbonic anhydrase activity of red blood cells from the volunteers. With minor modifications, he used the manometric method of Krebs and Roughton.[9]

Complete equipment for audiometry (soundproof room, audiometer, skilled operator) set up in connection with a study of the effect of noise on hearing made possible a study of the

Table 1.—Outline of Experiment

| Type of DDT | Group No. | Dosage of DDT | | Case No. | Number of Men | | | |
		Added (mg/Man/Day)	Total (Range, mg/kg/Day)*		Entering Experiment	Completing 21.5-Month Dosing Period	Completing 4-Year Project	Completing Additional Year
Technical	1	0.0	0.002-0.003	1-10	10	4	4†	2
	2	3.5	0.048-0.064	11-21	11	8	6	4
	3	35.0	0.374-0.585	22-32	11	7	6†	4
Recrystallized	4	35.0	0.447-0.550	33-43	11	10	8	6
Total					43	29	24	16

* The total dosage included dietary DDT as well as the administered dose. Dietary DDT, resulting from agricultural residues, was estimated at 0.184 mg per man per day on the basis of analyses of whole meals.[2] The total dosage varied somewhat according to the weight of the men.

† Included two Negroes.

effect of DDT on this function. The methods used, background information, and results of the study have been published.[10] Loss of hearing was expressed as percentage according to the method[11] current at the time.

Khairy[12] had found a change in the gait of rats fed DDT. For this reason, a study was made of the gait of each volunteer after dosing with DDT had been in effect for 18.8 months. Each man stepped into a shallow pan containing a little oil. He then walked naturally across the floor in such a way that the last part of his course was over a strip of brown wrapping paper rolled out like a carpet. The oil indicated the footsteps faintly and the record was made permanent by outlining the tracks and recording the man's name on the paper. Later, measurements were made of (a) the length of each step, (b) the width of each step (that is the perpendicular distance of the center point of each heel mark from a line drawn through the center of the immediately preceding and succeeding marks made by the opposite heel), and (c) the foot angle (acute angle between a line through the center of the heel and the center of the big toe and another line connecting the center of the heel print with the center of the next print of the same heel). The angle was recorded as negative for pigeon-toed people. The values were averaged for each man, and the ratio of his average width and length of step was calculated. The individual averages and ratios were then tabulated by group for statistical study.

Results

Effect on Health.—*Symptomatology and Physical Findings of Men Retained in the Study.*—Table 2 summarizes the condition of the volunteers at the time they entered the study. They varied in age from 24 to 49 years. All were in reasonably good health, although several showed borderline anemia, hypertension, or other abnormalities. Nearly all gave a history of prior disease, and some of their physical examinations revealed minor defects. A few gave a history of drug addiction. All had undergone hardship. Thus, the men probably were somewhat less sturdy on the average than men of the same age in the general population. The physical and laboratory findings for each volunteer in the predosing examination were essentially stable in his subsequent examinations. The same was less true of the subjective complaints encountered at each examination. Many of the complaints probably reflected either esprit de corps or an underlying personal anxiety rather than any organic change. For example, at an examination 50 days after the men began to take emulsion and 40 days after those in groups 2, 3, and 4 began to receive DDT, eight men in these groups reported their appetite had improved. The same was true of three men in the control group. Only one man said his appetite became worse (while he was receiving placebo), but it later increased above normal (a few days after he unknowingly began to receive DDT at the rate of 35 mg/day). This subjective improvement in appetite in some of the men was not accompanied by a corresponding change in weight. Three men (including one control) reported polyuria, but it was not confirmed by 24-hour urine samples. Complaints of twitching of muscles (most numerous in the control group) were not confirmed by examination. The reality of "kidney pain," "hunger pain," "heart burn," or "aching legs" could not be excluded by

83

Table 2.—*Condition of Volunteers at Beginning of the Study*

Group No.	1	2	3	4
No. of men	4	6	6	8
Age	28-38	24-45	26-49	25-43
(yr)	33.5 ± 2.1	33.5 ± 3.5	34.5 ± 3.1	33.9 ± 2.5
Height, cm	176.0-182.8	169.5-186.6	168.3-182.3	159.4-177.7
(in)	(69.3-72.0)	(66.7-73.5)	(66.5-73.8)	(62.8-70.0)
	179.5 ± 1.5	175.0 ± 2.5	176.6 ± 2.6	171.6 ± 2.1
	(70.2 ± 0.6)	(68.9 ± 1.0)	(69.5 ± 1.0)	(67.6 ± 0.8)
Weight, kg	71.2-88.9	56.7-78.0	60.3-93.8	62.6-83.4
(lb)	(157.0-196.0)	(125.0-172.0)	(133.0-207.0)	(138.0-184.0)
	78.7 ± 4.2	69.8 ± 3.8	74.0 ± 5.2	69.7 ± 2.4
	(173.5 ± 9.2)	(153.8 ± 8.3)	(163.1 ± 11.4)	(153.6 ± 5.4)
RBC (millions/cu mm)	4.8-5.1	5.1-5.5	4.3-5.7	4.7-5.3
	4.9 ± 0.1	5.3 ± 0.1	5.1 ± 0.2	5.0 ± 0.1
Hemoglobin	13.8-14.7	13.2-15.6	13.5-15.1	12.6-16.2
(gm/100 cc)	14.2 ± 0.2	14.4 ± 0.3	14.3 ± 0.3	14.1 ± 0.4
WBC (thousands/cu mm)	7.3-11.7	8.9-10.1	7.3-9.6	8.3-10.3
	8.9 ± 1.0	9.5 ± 0.2	8.3 ± 0.4	9.3 ± 0.3
Heart rate, rest	80-84	76-84	58-96	60-100
(beats per min)	83 ± 1.0	81 ± 1.3	79 ± 6.4	82 ± 4.0
Heart rate, exercise	104-128	104-170	104-142	100-136
(beats per min)	118 ± 5.0	126 ± 9.3	119 ± 5.4	116 ± 4.8
Heart rate, exercise-rest	68-80	60-112	48-94	56-96
(beats per min)	73 ± 2.5	80 ± 7.1	76 ± 7.2	82 ± 4.5
Systolic blood pressure,	116-132	104-130	102-140	108-132
rest (mm Hg)	125 ± 4.1	117 ± 3.6	122 ± 6.0	123 ± 3.0
Pulse pressure, rest	30-58	32-50	38-54	32-58
(mm Hg)	44 ± 5.8	41 ± 42	46 ± 2.8	44 ± 3.3
Systolic pressure,	124-132	102-140	108-160	108-142
exercise (mm Hg)	129 ± 1.7	127 ± 5.2	134 ± 9.3	125 ± 40
Pulse pressure,	30-58	38-62	36-80	42-66
exercise (mm Hg)	44 ± 5.9	50 ± 4.0	52 ± 6.7	50 ± 3.5

Table 3.—*Range, Mean and Standard Error of the Concentration of DDT (ppm) in Body Fat of Volunteers*

Group	1	2	3	4
Dosage (mg/man/day)	0	3.5	35	35
Type of DDT	Technical	Technical	Technical	Recrystallized
No. of men	4	6	6	8
Time				
Before exposure	2.6-5.8	3.1-3.9	3.5-5.2	2.0-12.5
	4.3 ± 0.83	3.4 ± 0.2	4.1 ± 0.3	9.0 ± 1.2
12.2 mo exposure	7-15	22-45	173-246	122-307
	10.3 ± 1.7	32.2 ± 3.3	201.2 ± 13.3	211.0 ± 22.8
18.8 mo exposure	5-22	22-77	104-276	166-508
	16.3 ± 3.9	49.2 ± 7.7	205.0 ± 27.7	268.6 ± 44.9
21.5 mo exposure	16-30	39-76	105-619	129-657
	22.0 ± 2.9	50.2 ± 5.5	280.5 ± 79.5	325.0 ± 62.0
26.4 mo in experiment	8-17	14-47	50-216	83-390
4.9 mo recovery	13.3 ± 2.7	31.4 ± 5.3	124.2 ± 23.1	160.8 ± 33.9
33.0 mo in experiment	10-28	14-46	60-313	101-376
11.5 mo recovery	19.0 ± 9.0	26.3 ± 4.6	139.8 ± 38.1	235.5 ± 33.4
39.5 mo in experiment	12-24	25-47	29-330	76-232
18.0 mo recovery	18.0 ± 6.0	36.8 ± 3.8	126.7 ± 45.4	156.4 ± 23.0
47.0 mo in experiment	20-21	10-55	33-253	48-214
25.5 mo recovery	20.5 ± 0.5	33.2 ± 7.0	99.8 ± 34.8	105.1 ± 20.1
59.3 mo in experiment	6-16*	5-19*	24-102*	20-109*
37.8 mo recovery	11.0 ± 5.0	14.0 ± 3.2	56.8 ± 19.1	46.0 ± 13.4

* The number of men in groups 1 to 4 at this time was two, four, four, and six, respectively.

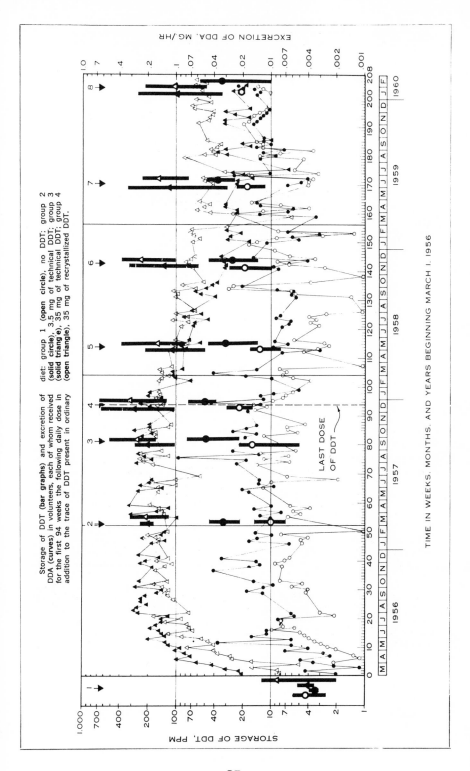

Storage of DDT (**bar graphs**) and excretion of DDA (**curves**) in volunteers, each of whom received for the first 94 weeks the following daily dose in addition to the trace of DDT present in ordinary diet: group 1 (**open circle**), no DDT; group 2 (**solid circle**), 3.5 mg of technical DDT; group 3 (**solid triangle**), 35 mg of technical DDT; group 4 (**open triangle**), 35 mg of recrystallized DDT.

EXCRETION OF DDA, MG./HR

STORAGE OF DDT, PPM

TIME IN WEEKS, MONTHS, AND YEARS BEGINNING MARCH 1, 1956

LAST DOSE OF DDT

Table 4.—DDT In Fat of People With Ordinary Diet and No Occupational Exposure*

Range (ppm DDT)	Cases, Percent				
	Laug et al[15]	Hayes et al[1]	Hayes et al[16]		This Study
			Group 3	Group 4	
0.0	20	0	3	1	0
0.1-1.0	9	0	0	1	0
1.1-5.0	28	25	69	45	58
5.1-10.0	28	61	25	45	28
10.1-20.0	12	14	3	8	14
Over 20.0	3	0	0	0	0

* During the period 1951 to 1958.

Table 5.—Significance of Differences in DDT Storage in Fat Associated With Dosage, Exposure, and Type of DDT

Condition	Group	Comparison	Significance
Type of DDT		Dosage (mg/man/day)	
Technical for	1 vs 2	0vs3.5	<0.005
21.5 mo	2 vs 3	3.5 vs 35.0	<0.025
Type of DDT		Duration of Exposure (mo)	
Technical	1	21.5 vs 12.2	<0.025
		21.5 vs 18.8	>0.20
Technical	2	21.5 vs 12.2	<0.025
		21.5 vs 18.8	>0.90
Technical	3	21.5 vs 12.2	>0.20
		21.5 vs 18.8	>0.20
Recrystallized	4	21.5 vs 12.2	>0.05
		21.5 vs 18.8	>0.40
Duration of Exposure (mo)		Type of DDT	
12.2	3 vs 4	Technical vs recrystallized	>0.70
18.8	3 vs 4	Technical vs recrystallized	>0.20
21.5	3 vs 4	Technical vs recrystallized	>0.60

examination but did not correspond with any physical or laboratory finding. On the contrary, complaints of dermatitis usually were easily confirmed by examination, but like the other complaints were at least as frequent in the controls as in the men who received DDT.

The men showed a slight tendency to weight loss during the study. Among those who finished the study, the average loss at the end of the dosing period was 0.5, 2.0, 0.5, and 0.2 kg (1.2, 4.5, 1.2, and 0.5 lb) for groups 1, 2, 3, and 4, respectively. The situation at the end of the four-year study was essentially unchanged. During the entire period, the largest loss (6.8 kg [15.0 lb]) and the largest gain (12.3 kg [27.1 lb]) both oc-

curred in group 3, but almost equal changes (−5.9 kg [−13.0 lb] and +6.8 kg [+15.0 lb]) occurred in the control group.

Because DDT in sufficient dosage produces tremor and incoordination, particular attention was paid to the neurological tests. At each examination the following were recorded: gait, including ability to walk on toes and heels; coordination, including Romberg, finger to nose, heel to shin, adiadochokinesia, rebound, past-pointing, tremor, handwriting, and "tongue twisters"; deep reflexes, superficial reflexes, and clonus; muscle strength, including all major movements of the neck, shoulder, elbow, wrist, fingers, hip, knee, and ankle; sensation, including light touch, pinprick, two-point discrimination,

		Number of Doses Missed	
Group	Dosage Added (mg/Man/Day)	Range	Mean
1	0.0	7-30	19.0
2	3.5	7-68	28.7
3	35.0	6-24	12.8
4	35.0	8-72	30.4

Table 6.—Recorded Failure of 24 Volunteers to Take Some Portion of 656 Intended Doses

position sense, vibration sense, stereognosis, and differentiation of heat and cold; and function of the cranial nerves, including identification of an odor, visual fields, reactivity of pupils, extraocular movements, strength of bite, sensation of the face, coordination and strength of facial muscles, hearing, pharyngeal reflex and voice, and protrusion of the tongue. Although the volunteers varied in their ability to perform some of the tests, especially the "tongue twisters," there was no relation between coordination, tremor, or the performance of any other test and intake of DDT.

With the two exceptions described in the following paragraphs, there was no essential difference in the medical history, systems review, or physical findings of men who completed the study and those who did not.

Occurrence of Illness.—No illness that could be attributed to DDT was seen during the study. The men who quit of their own accord did not complain of illness and there were no significant findings on examination. As it happened, most of them had received no DDT.

Two men (not included in the statistical tables) became severely sick during the project. One of them (case 22) suffered a myocardial infarction after he had received 597 doses of technical DDT at the rate of 35 mg daily. He was then 42 years old. DDT dosage was discontinued but additional urine samples were collected for study. The patient had a stormy course and did not recover completely. He was able to return to light duty. At the last examination, when he was 45 years old, there were no significant findings except those referable to the circulatory system. At that time, a diagnosis was made of arteriosclerotic heart disease with anginal syndrome. Because of his susceptibility to angina, an exercise tolerance test was not done. At rest, the pulse rate was 74 beats

per minute and the blood pressure was 102/68 mm Hg.

The second man who became sick (case 41) was recovering from a gallbladder operation when he entered the study, and he was taken only because of his own insistence. (In the earlier study, men with a history of jaundice had shown no ill effects following dosage with DDT.) Patient 41 received recrystallized DDT at the rate of 35 mg daily. He became sick after he had been in the project less than five months. He was 44 years old at the time. He experienced increasing anorexia and nausea for four to five days before he was hospitalized. On admission, he had generalized malaise but denied pain. He had slight jaundice, low-grade fever, no abdominal tenderness, and no hepatomegaly. The icterus index was elevated, the urine showed free bile, and the sulfobromophthalein retention was 70% in five minutes and 30% in 30 minutes. Treatment consisted of complete bed rest, multivitamins, and a diet high in protein and carbohydrate but low in fat. Dosing with DDT was stopped after a total of 133 doses and was not resumed. It was concluded that the jaundice was not caused by obstruction, but the cause of the hepatitis was not established. A viral origin was not ruled out. Recovery occurred gradually. By the time of the next regular examination (one year after beginning the project and seven months after onset), the volunteer was working regularly and there were no abnormal physical or laboratory results related to the liver; sulfobromophthalein retention, plasma cholinesterase, and SGOT levels were normal. From this time on he showed no evidence of liver disease. However, he had a number of complaints not recorded at the first examination. The complaints included "low blood pressure" (actual blood pressure, 96/64 mm Hg), frequent difficulty in breathing, bloated feeling after eating, bleeding hemorrhoids, pain in the shoulders, pain in the right hip and back radiating to the right testis, burning pain on urination, exhaustion in the morning, difficulty in going to sleep, and generally poor health. Physical, neurological, and laboratory findings were essentially normal except for hemorrhoids.

The volunteer received a hemorrhoidectomy. At other times he received sitz baths, heat treatments, and prostatic massage. Van Buren sounds up to 28 F were passed to correct difficulty dating back to an old case of gonorrhea.

In spite of treatment, the patient continued to complain, and after the hemorrhoidectomy he blamed some of his pains on the spinal anesthesia. From 1957 onward, he had more varied and more severe complaints than could be accounted for by examination, including at different times examination for kidney stone and stomach ulcer. In spite of his varied difficulties, he stated at the last examination (February 1961) that he felt the project had not had any effect on his health. At no time was there any objective finding to indicate a relationship between illness and DDT storage.

White Blood Cells.—Three of the men in group 4 showed 9% to 12% eosinophils at a single examination. All other values for eosinophils were normal.

Of the 24 men who completed the study, ten (42%) showed a reversal of the granulocyte-lymphocyte ratio; five of them (21%) showed the change before they received any dosing with DDT. The condition tended to be typical of the individual but was not constant; several volunteers showed a reversal of the ratio more than once but none showed it at every examination. The frequency of the finding in those who received added DDT (3.5 or 35 mg) was actually less at each examination during the period of dosing and also during the recovery period than it was in the same men before dosing. Reversal of the ratio occurred most often among the control group. The lowest observed percentage of total granulocytes was 35 and this occurred in the control group. The lowest total number of granulocytes was 2,244/cu ml, associated with a percentage of 44, which occurred in case 14, 37.8 months after his last dose at the rate of 3.5 mg daily. Before dosing, the same men had 43% total granulocytes. The total number of granulocytes fell below 2,500/cu ml in only the one case (0.74% of the samples).

Red Blood Cells and Hemoglobin Level. —There was no indication that exposure to DDT had any effect on the red blood cell count (RBC). This was true even though some degree of anemia (less than 4.5×10^6 cells/cu mm) was found in different groups in 6% to 17% of examinations. The proportion in the control group was 17%. The lowest count found was 3.7×10^6 cells/cu mm in a man in group 4, all of whose other values were normal. The lowest value in the control group was 3.9×10^6 cells/cu mm in a man who was anemic on other occasions also. Four men on one occasion each had counts of 6.1 to 7.8×10^6 cells/cu mm. All were receiving DDT but all reverted to normal, two of them while still taking DDT and two of them at the first examination after regular dosing was stopped.

Hemoglobin level was entirely unaffected by exposure to DDT. Nearly all the values were within normal limit. The lowest observed was 10.5 gm/100 cc. In nearly every instance, low hemoglobin value was associated with a low RBC, but many of the low counts were accompanied by acceptable concentrations of hemoglobin. Abnormal values (below 13 gm/100 cc) were more frequent in the controls than in the other groups. The six (out of 60) values less than 13 gm/100 cc that occurred in men fed DDT during the dosing period returned to normal during subsequent examinations, and four of them returned to normal while DDT was still being administered.

Cardiovascular Status.—As shown in Table 2, resting heart rates as high as 100 beats per minute were encountered in men who completed the study before they received any DDT. A resting rate of 118 beats per minute was found in a control who quit. There was a tendency for the resting heart rate to be slightly lower after the preexposure examination, perhaps reflecting a decrease in anxiety after the study got under way. This was true for those receiving DDT and also for controls.

The pulse rates during exercise were not remarkable. Counting all the volunteers, the heart rate two minutes after exercise exceeded 99 beats per minute, twice as often as at rest. There was no correlation with exposure to DDT.

There also was no correlation between blood pressure and exposure to DDT. As shown by Table 2, some of the men had mild or threshold hypertension when they began the study. In all, there were seven resting values of 140 mm Hg or greater in men who completed the study and one such value in a control who quit. In every instance, these readings were taken before DDT was administered or in a man who had a reading of 130 mm Hg or more before dosing began. The same tendency to hypertension was seen to a more marked degree following exercise, but again there was no indication that DDT was involved. Of 26 readings of 140 mm Hg or higher in the entire study, 11 were recorded before DDT was given. There was no essential difference in the cardiovascular performance of the men who finished the study and those who did not.

In summary, no serious cardiovascular dysfunction was found among men who completed the study. The cardiovascular status of several of the volunteers was less than ideal, probably reflecting their age, lack of regular exercise, and perhaps past conditions of life. There were few athletes among them. However, the mild dysfunction observed showed no relation to the amount of DDT ingested.

Liver Function Tests.—Plasma cholinesterase activity was measured before dosing started and 12.2 and 18.8 months after it began. The range for all determinations were 0.47 to 1.64 Δ pH/hr, almost identical to the range found by Rider et al[13] in a study of 400 normal men. In the present study, there was no significant change in plasma cholinesterase activity either as a function of time or DDT intake.

SGOT was measured twice, once 12.2 and once 18.8 months after dosing began. The normal activity was and still is regarded as 12 to 40 Karmen units/ml. However, Dr. Schneider's laboratory found that there is a second small distribution within the range of 40 to 70 Karmen units/ml among people presumed to be completely normal. Values within this higher range and a single value of 119 Karmen units/ml were found in 12% of determinations in the first series and in 9% in the second series. Since all of the values above 40 Karmen units/ml were for men receiving some DDT and since there was no measurement of SGOT activity before dosing with DDT began, no final conclusion is possible. However, the average values for all groups were normal; the highest value encountered was in a man who received DDT at a rate of only 3.5 mg/kg/day and

whose second value was 25.3 Karmen units/ml; and the highest value for any man who had received technical DDT for 18.8 months was 37 Karmen units/100 ml. Because SGOT values did not correspond to DDT dosage levels, there is no evidence that DDT had any influence on SGOT.

Sulfobromophthalein retention was measured before dosing started and 12.2 and 18.6 months after it began. No abnormal values appeared in the initial test. There were two high values (11% and 12%) in the second test and five high values (6% to 12%) in the third test. All of these high values appeared in groups 3 and 4 except one (6%) which occurred in a control. Two men in group 4 had one or more tests with a sulfobromophthalein retention greater than 5% and two SGOT values greater than 40 Karmen units/ml. Other increases in sulfobromophthalein retention did not appear in men who at any time had an SGOT value over 40 Karmen units/ml.

Carbonic Anhydrase Activity.—The activity of carbonic anhydrase in RBC taken from the volunteers at the last regular examination (47 months in the study) was statistically indistinguishable from that of controls in the study and other controls not connected with the study. Since the volunteers in groups 2 to 4 had their last dose of DDT 25.5 months before the enzyme assay, it is of particular interest that tests were also normal for blood from men working in a DDT formulating plant and who had worked there for an average of 10.9 years.

Hearing.—Some of the volunteers had essentially perfect hearing when they entered the study and others had defective hearing. The largest combined hearing loss measured was 26% in a man whose preexposure loss was 25%. Each person retained his characteristic pattern throughout the study. There was, to be sure, some small random variation in some individuals from one measurement to another; the greatest worsening observed for any individual for any measurement compared to his first was 8%, and the greatest apparent improvement was also 8%. The greatest average change in either direction at the end of the dosing period was an apparent improvement of 1.6% in the score of group 4 receiving recrystallized DDT at a rate of 35 mg per man per day. At the end of the

| Type of DDT | Added Dosage (mg/Man/Day) | Group* | Concentration of DDT: Range, Mean, and SE | | Significance of Difference (P) |
			Present Study, 21.5 mo (ppm)	Earlier Study, 11 mo or More (ppm)	
Technical	0.0	1	16-30 (22.0 ± 2.9)	8-17 (12.5 ± 4.5)	>0.10
	3.5	2	39-76 (50.2 ± 5.6)	26-33 (29.8 ± 1.4)	<0.025
	35.0	3	105-619 (281.0 ± 79.5)	101-367 (234.0 ± 21.4)	>0.40
Recrystallized	35.0	4	129-659 (325.0 ± 62.2)	216-466 (340.0 ± 36.4)	>0.20

* Present study only.

study, the first three groups, including the control, showed a slight worsening (maximal average, 1.2%), but group 4 still showed an apparent improvement (0.8%). Of course, hearing tends to become defective with advancing age, but in the volunteers this expected change was very small during the four years of study. Repeated intake of DDT had no detectable influence on hearing. This is consistent with the negative results of single large doses either in volunteers or in persons accidentally poisoned by the compound.[14]

Gait.—Study of the men's gait revealed no trend related to DDT dosage. The work of Khairy[12] had revealed that the ratio of width to length of step decreased in rats at a dietary level of 100 ppm (about 5 mg/kg/day) but increased above normal in rats receiving 200 to 600 ppm of DDT in their feed. Failure to find any effect in man may be related to dosage, which varied from 0.374 to 0.585 mg/kg/day in groups 3 and 4, but was much higher in the rats.

There was some tendency for tall men to take long steps. The width of step showed no detectable relation to height or anything else. Thus the ratio of width to length tended to vary inversely with height; the coefficient of correlation ranging from essentially 0 to −0.83 for different groups.

Storage of DDT-Derived Materials.—*Storage of DDT in Men Who Completed the Study.*—The storage of DDT is shown in Table 3 and is represented by bar graphs in the Figure. The storage of DDE in the volunteers is shown in a later table. The amount of DDT storage found before dosing began was similar to that found in other sets of samples from the general population at about the same period (Table 4).

In group 2, who received technical DDT at the rate of 3.5 mg per man per day, there was a statistically significant increase in the storage of DDT beyond what they reached in 12.2 months. The same was true of the controls but not of the groups receiving 35 mg/kg/day (Table 5). All of the groups showed some increase in average storage after 21.5 months of dosing as compared to 18.8 months, but the difference was not statistically significant in any instance.

The government-employed technician assigned to prepare and label a dose for each volunteer each day also had as his duty to record any dose unclaimed at the end of the day (proving failure of a particular volunteer) or to record the fact if a volunteer claimed a dose already issued. The technician acknowledged those latter errors as the result of his own carelessness. The men whose results are shown in the Figure were recorded as missing 7 to 72 doses each, as shown in detail in Table 6. Furthermore, there was objective evidence from the dosage record that men in the control group occasionally got doses containing DDT intended for men in other groups. This is borne out by the analytical results showing an increase of storage in the control group. The number of failures and mistakes may have been greater if the technician failed to record some of them. It simply was not possible to ensure as accurate a distribution of doses as had been done in the earlier study. In spite of the recognized difficulties, all of the results for storage and excretion were proportional to intended dosage (Figure and Table 5). The irregularity of dosage may help to account for the failure of the groups in the present study to approach a steady state of DDT storage within a year. Evidence to this effect is the negative correlation between the num-

Table 8.—Range, Mean, and SE of the Concentration of DDE (ppm)
in Body Fat of Volunteers

Group	1	2	3	4
Dosage (mg/man/day)	0	3.5	35	35
Type of DDT	Technical	Technical	Technical	Recrystallized
No. of men	4	6	6	8
Time				
Before exposure	3.4-17.0 8.3±3.0	5.5-8.2 7.0±0.8	4.2-10.9 7.0±0.9	2.6-21.7 12.5±2.1
12.2 mo exposure	5-7 6.0±0.6	4-8 6.2±0.8	0-36 6.0±6.0	20-66 37.8±6.2
18.8 mo exposure	2-10 7.0±1.8	6-18 11.2±2.0	15-30 26.2±2.3	18-101 58.5±10.1
21.5 mo exposure	4-11 7.0±1.5	12-30 17.0±2.8	16-75 35.8±8.6	51-142 92.1±15.6
26.4 mo in experiment 4.9 mo recovery	2-12 7.3±2.9	4-20 12.5±2.4	14-92 43.0±11.4	38-169 82.9±17.9
33.0 mo in experiment 11.5 mo recovery	6-18 12.0±6.0	4-25 13.2±3.0	24-64 39.3±6.1	36-155 92.8±13.6
39.5 mo in experiment 18.0 mo recovery	9-30 19.5±10.5	11-55 28.0±7.4	34-94 51.3±9.8	57-160 107.9±14.3
47.0 mo in experiment 25.5 mo recovery	19-29 24.0±5.0	15-34 22.0±3.0	29-113 56.3±12.9	37-214 99.1±20.8
59.3 mo in experiment 37.8 mo recovery	7-35* 21.0±14.0	9-24* 16.0±3.4	25-81* 50.8±11.5	29-102* 68.3±10.3

* The number of men in groups 1 to 4 at this time was two, four, four, and six, respectively.

ber of doses known to have been missed and the concentration of DDT stored at 12.2 months for men on the high dosage levels (r = 0.69 for group 3 and r = 0.35 for group 4). There was no correlation between storage and recorded failure to take doses in group 2 (r = + 0.11), but this lack of correlation probably can be accounted for by the fact that taking the wrong dose would be relatively much more important for this group than merely missing a dose.

It is probable that a steady state was approached by 18.8 months in the present study; the values were not statistically different from those taken 2.7 months later (Table 5). Furthermore, in three of four instances in which a comparison was possible, the storage of DDT was not significantly higher in the present study after 21.5 months than it was in comparable groups of the earlier study after 11 months or more (Table 7). The fact that men receiving daily doses of 3.5 mg or less stored more—and, in one instance, significantly more—in the present study than did men in the earlier study may be correlated with the fact that the men in the present study probably occasionally took doses at the rate of 35 mg per man per day by mistake.

The average storage of recrystallized DDT was always slightly higher than that of technical DDT but the difference did not reach a level of statistical significance for sets of samples taken at the same time (Table 5).

Four Negroes entered the experiment, two in group 1 and two in group 3. None of these men had an unusual amount of DDT storage before dosing began. The two who received DDT did not store it to an unusual degree during the dosing period as compared to other members of their group. However, one had higher storage than any other man in the group at every examination during the recovery period. The other Negro was second highest in his group in half of the examinations during recovery.

Loss of Stored DDT in Men Who Completed the Study.—Storage loss of DDT occurred slowly after dosing was discontinued. After 25.5 months for recovery, storage of DDT in the two groups on 35 mg per man per day was reduced to 32% and 35% of the values they reached at the end of the dosing period. During the same recovery period, storage of DDT was reduced only to 66% in the group on 3.5 mg per man per day and to 93% in the controls. After the planned study was completed, 16 men stayed on for further investigation. Their storage of DDT 37.8 months after last dosing was 14%, 20%,

Group	1	2	3	4
Dosage (mg/man/day)	0	3.5	35	35
Type of DDT	Technical	Technical	Technical	Recrystallized
Duration of Study (mo)	Proportion of DDE in Total (%)			
0	66	67	63	58
Dosage 12.2	37	16	8	15
period 18.8	30	19	11	20
21.5	24	27	11	22
26.4	35	32	26	34
Recovery 33.0	39	33	22	28
period 39.5	52	43	29	41
47.0	54	40	36	49
59.3	66	53	47	60

26%, and 50%, respectively, based on their personal maximal values at 21.5 months and listed in the same order as the group percentages above. The values are consistent in indicating proportionately slower loss of DDT at lower storage levels.

Storage of DDE in Men Who Completed the Study.—Storage of DDE increased throughout the dosing period in the two groups on 35 mg per man per day and even continued to increase slightly after dosing was stopped (Table 8). In the group on 3.5 mg, an increase was first detected in the samples taken at 18.8 months but there was again some further increase after dosing stopped. In the control group, the increase in DDE storage was postponed until nearly a year after the last dose for the other volunteers. The increase in DDE storage observed in all groups sometime after dosing was stopped suggests that, in man, stored DDT is gradually metabolized in part to DDE which is stored more tenaciously in the fat than DDT itself is stored. The fact that the storage pattern for DDE does not parallel that for DDT is illustrated by Table 9. The delay in DDE storage following DDT dosing suggests that man does not convert a high percentage of DDT to DDE in the process of absorption from the alimentary tract as occurs in the rat.[17]

Men who stayed on for an additional year showed a slight decrease in the concentration of DDE in their fat during this period. Because the storage of DDT decreased faster, the percentage present in the form of DDE actually increased.

Storage in Men Who Failed to Complete the Study.—The man (case 41) whose dos-

age was discontinued after only 133 doses had DDT in a concentration of 71 ppm at the time the 12.2-month biopsy specimen was taken. Thereafter, his storage decreased very slowly, the last value during the regular study being 27 ppm. Because of the difference in his dosing schedule, only his predosing value resembled those of his group. With this exception, there was no meaningful difference in storage of DDT or DDE by men who failed to complete the study as compared to those who did finish. The other man, whose dosing was discontinued and who continued to give samples (case 22), missed less than two months of dosing, and his values were indistinguishable from those of his group. Out of 63 possible comparisons (excluding case 41), only five DDT values for men who failed to finish were either smaller or larger than corresponding values for those who did finish. There was no trend; and in all three instances in which later measurements were made, the later values for those who had been out of line, fell within the range of men in the same group who finished the study.

The situation was entirely similar for DDE storage. Here, too (with the exception of case 41), there were only five values outside the range seen for the 24 volunteers, and one of these involved a control.

Excretion of DDT-Derived Material.—*Excretion in Men Who Completed the Study.* —Urinary excretion of DDA was 4 to 15 times greater in the first samples from men receiving DDT at a rate of 35 mg per man per day only four to five days after they began dosing than it was in the controls at the same time. Excretion in the high dosage

Table 10.—Urinary Excretion of DDA by Men in Group 3 During the 38th Through the 94th Week of Dosing*

Case No.	DDA Excretion (mg/hr)	
	Range	Mean ± SE
24	0.05-0.16	0.089 ± 0.006
27	0.04-0.20	0.095 ± 0.011
26	0.08-0.37	0.194 ± 0.020
28	0.02-0.36	0.233 ± 0.020
30	0.11-0.54	0.262 ± 0.023
31	0.15-0.43	0.269 ± 0.014

* Cases arranged in order of increasing mean excretory values.

groups increased rapidly at first and then more gradually (Figure). About half the maximal rate was achieved in two to three months and there was no evidence of further increase after six to eight months in any group. After dosing with DDT was stopped, excretion of DDA decreased gradually in the groups that had received 35 mg per man per day. However, the concentration of DDA in the urine of these two groups was still above control values 25.5 months after the last dose.

During approximately the first eight months after the last dose, excretion appeared to decrease gradually in the group on 3.5 mg per man per day and also in the control group. There was then a gradual increase in the excretion values for both groups, and this increase persisted for about 21 months. The results for DDT storage appear consistent with this trend. The average differences in excretory rate were small, being in the order of 0.008 mg/hr at most. The increase, though small, suggests the unintentional introduction in August or September 1958 of a new source of DDT intake at a rate of about 1 to 2 mg per man per day. However, no such source was identified and mention of it is entirely speculative. Whatever its cause the change did not involve an annual cycle, since the apparent increase continued for about 21 months.

Variation between individuals in the excretion of DDA was marked. This variation is illustrated by Table 10, which gives an analysis of the results for group 3. Obviously, there was no statistical difference between the rates of certain men. However, the rate for case 26, which fell about midway between the extremes, was significantly different from both the high and the low extremes ($P < 0.05$ and < 0.001, respectively and the difference between the extremes (cases 24 and 31) was highly significant ($P < 0.001$). In fact, the highest rate ever reached in case 24 was almost the same as the lowest rate shown in case 31.

Cases 24 and 26 involved Negroes. Because the total number of men in the group is small, no conclusion can be drawn about whether race or merely individual difference influenced the position of cases 24 and 26 in Table 10.

Although a particular level of excretion was characteristic of each individual, there were also a few instances (1% or less of samples, depending on the group) of striking increases or decreases in excretion rate. The degree of increase varied from two to ten or more times. Some of these changes were detected in a single sample only, and in these instances, there was no way to exclude entirely the possibility of laboratory error. However, the fact that the change occasionally did persist in two or three consecutive, approximately weekly samples from the same person suggests that the striking changes were real and due to some unidentified, relatively temporary physiological change. This was made almost certain by finding that, on different occasions, series of values about twice the usual continued for 5 to 11 consecutive days in one man who contributed daily samples. Some of the high values in different volunteers were associated with high urinary output but this relationship was far from consistent.

At a dosage of 35 mg per man per day and during the period from 38 weeks until the end of dosing at 94 weeks, the concentration of DDA associated with technical DDT varied from 0.18 to 9.21 ppm and averaged 2.98 ppm. The corresponding values for recrystallized DDT ranged from 0.40 to 6.27 ppm and averaged 1.88 ppm.

After a steady state was reached (about 38 weeks), the daily excretion of DDA associated with technical DDT at 35 mg per man per day averaged 5.7 mg (expressed as DDT) for the entire group, while that for recrystallized DDT administered at the same rate averaged 4.7 mg. Thus, the average proportion of the dosage accounted for by excretion was 16% and 13%, respectively. The excretion associated with technical DDT at a dosage of 3.5 mg per man per day averaged 0.46 mg daily or 13% of the daily dose.

Relation of Excretion and Storage of Recrystallized and Technical DDT.—In the earlier study, it was found that recrystallized DDT was stored at a statistically significant higher rate than technical DDT given at the same dosage. The storage of recrystallized DDT was greater in the present study also, but not to a statistically significant degree (Table 5). However, the existence of a real difference was suggested in the present study not only by the minor difference in storage but also by the excretion patterns associated with the two formulations. It might be concluded from the Figure that more recrystallized DDT was stored during the dosing period because less was excreted. The same reasoning accounts for the curious crossing over of the excretion curves for group 3 and 4 just a few weeks after dosing was stopped. The more efficient excretion of the technical DDT soon reduced its storage so much that its absolute excretion dropped below that of the recrystallized material. However, the difference in storage, which according to this hypothesis should have increased, actually became less in 1959 and 1960 and reversed its direction in 1961. It appears that any difference that may exist in the storage and excretion of technical DDT as compared to recrystallized DDT is complex and not understood at this time.

Men who continued in the study for an additional year showed a continued gradual decrease in the excretion of urinary DDA. Toward the end of the year, no DDA could be detected in many samples from men who had received no DDT or 3.5 mg per man per day. During the last three weeks, the weekly average was 0.006 to 0.009 ppm for men who had received technical DDT at 35 mg per man per day and 0.016 to 0.019 ppm for those who had received recrystallized DDT at the same rate.

Urinary Excretion in Men Who Failed to Complete the Study.—Excretion of DDA never became very high and began to decrease promptly when dosing was stopped after only 133 doses in one man (case 41). Thus, nearly all his excretory levels were atypical of his group.

In case 22, dosing was continued almost the full intended time, and excretion of DDA was similar to those of others in his group at the time he stopped taking DDT. Taking into account the difference in his dosage schedule, the rate at which his excretion decreased was generally similar to other rates for his group. However, he was unusual in having a period of relatively very high excretion of DDA (0.12 to 0.21 mg/hr) beginning 10.5 months after dosing and lasting two months. He showed somewhat high excretion for the next two months also and then became similar once again to the rest of his group.

There were two other instances of sudden, temporary increases in excretion among men who did not complete the study (and three instances among men who did complete it).

There was no essential difference in the excretion of DDA among men who completed the study and those who did not.

Interference in the Measurement of DDA.—In the course of following case 41 through his illness, some strong yellow blanks were encountered in the analysis of urine samples for DDA. The yellow color resulted in falsely high values for DDA if the samples were read in the usual way. A review of medications suggested that sulfisoxazole might be the cause of the yellow color. This was proved by analyzing urine samples from a man not connected with the project both before and after he took sulfisoxazole at the usual therapeutic dosage. The color-producing material appeared in high concentration, in the urine within 1.5 hours. The interference from sulfisoxazole in the method of Cueto et al[3] was significantly reduced by washing the benzene extract with water prior to nitration until a wash of neutral pH was obtained. It is possible that some other drugs may interfere with the colorimetric measurement of DDA in urine.

Failure to Detect Fecal Excretion.—Detection of DDT-derived material in the feces was attempted without success. Because less than 20% of the dosage was recovered, it seems likely that a major part of the total excretion involves polar conjugates of DDT metabolites that are not extractable or metabolites that do not give a color in the Schechter-Haller[18] reaction.

Comment

Methods of Chemical Analysis.—The modern method of analyzing chlorinated hydrocarbon insecticides by gas chromatography permits separate measurement of different isomers and metabolites in tissues and permits a distinction between DDE and DDA in urine. This latter distinction, particularly, would have been useful in the present study. However, the results reported are considered valid. When the same tissue samples or tissue extracts were analyzed by the Schechter-Haller[18] method and by gas chromatography, the results for total DDT-derived material were either indistinguishable[19] or small.[20] In men with occupational exposure to DDT, the two methods of analysis gave indistinguishable results for the excretion of DDA.[21]

Laboratory Findings.—Ortelee[22] reported an eosinophil count of 7% or more in three men and a reversal of the granulocyte-lymphocyte ratio in two men out of 40 with occupational exposure to DDT. The men showing these changes had been exposed for 1.4 to 6 years. The total frequencies of granulocytes in those showing the reversal of ratio were 1,908 and 2,184 per cubic milliliter, respectively. Ortelee[22] drew no conclusion about the relationship between exposure and the changes observed. Similar results in men exposed to DDT were reported by Laws et al,[21] who considered the variation normal. It is clear from the results given above that the similar changes observed in the present study showed no relationship to the dosages of DDT administered. It is open to question whether the observed reversal has any significance at all. Booth and Hancock[23] found that 14.4% of the normal males whom they studied showed reversal of the granulocyte-lymphocyte ratio in one or more counts over a period of two years. The tendency to reversal of ratio, like the total white blood cell count, was characteristic of the individual. Some of their subjects showed a reversal of ratio in a third or more of their counts. In the study by Booth and Hancock,[23] 0.4% of the total granulocyte counts were below 2,000/cu ml and 1.8% were below 2,500/cu ml. A reversal of the granulocyte-lymphocyte ratio was seen more frequently in this study than in the study by Booth and Hancock,[23] but total counts below 2,500/cu ml were not

any more frequent than they were in normal people in England. Furthermore, reversal of the granulocyte-lymphocyte ratio was observed in 22% of 269 normal adults studied before DDT was introduced.[24]

The only clinical laboratory results that might have been influenced by DDT dosage were SGOT activity and sulfobromophthalein retention, and even the minor changes in them were not dose-related or statistically significant.

Measurement of RBC carbonic anhydrase was considered of interest because dogs that received DDT at a high rate excreted more urine than their controls,[25] and the carbonic anhydrase of RBC is inhibited in vitro by DDT.[26] Since the work of Davenport,[27] it generally is thought that carbonic anhydrase is involved in kidney function, and acetazolamide, a potent inhibitor of the enzyme, is now used commonly as a diuretic in therapy. It is possible that inhibition of the enzyme was responsible for the increased production of urine by dogs in the early experiment, but this remains to be proved. The dogs ingested DDT at the rate of 100 mg/kg/day, while the maximal rate of intake for any volunteer was only 0.585 mg/kg/day. In view of the negative results in volunteers and formulating plant workers, it appears unlikely that any dosage of DDT encountered by people (except perhaps in connection with accidental ingestion or suicide) will have any effect on the activity of carbonic anhydrase.

Storage and Excretion of DDT-Derived Material.—*Dynamics of Storage.*—The storage of DDT continued to increase during the entire dosing period. However, the increase was not statistically significant in any group after 18.8 months and only in the control and the low dosage group after 12.2 months. The concentrations of stored DDT reached in 94 weeks in this study by the groups receiving 35 mg per man per day were statistically indistinguishable from those reached in about a year by corresponding groups in an earlier study. In this study and the earlier one, the form of the storage curves and the lack of statistical difference between succeeding values from the same men offer reason to conclude that the final levels in each instance represent a steady state of storage or at least approach such a state. The results are consistent with the model proposed by

Robinson[28,29] for exponential approach to storage equilibrium. Graphic analysis (not shown) based on this concept suggests that the values finally reached in the present study may have been in the order of 85% to 90% of equilibrium. Data for the previous study do not exclude this possibility in that instance.

There is no doubt that storage increased more slowly in the present study, and it appears likely that irregularity of dosing was at least partially responsible. Unfortunately, the introduction of this unintentional variable limits precision. Any future attempt to study the kinetics of DDT storage in man should involve dosing for three years in order to permit multiple samples within the range of 85% to 90% of true equilibrium and final achievement of a 95% to 99% approach to true equilibrium even in the face of irregular dosing. Furthermore, every effort should be made to achieve a regular daily schedule of dosing. This would permit testing the hypothesis that storage is 85% to 90% complete in 12 months as demonstrated by repeated samples within the range of 95% to 99% of true equilibrium during the last year of study.

Relation of Storage and Excretion.—The rate of excretion of DDA showed no further increase after about 38 weeks of dosing. The explanation for the apparent difference in the timing of the attainment of equilibrium of storage of DDT and the excretion of DDA has not been found.

The man (case 24) who showed the slowest excretion of DDA (Table 10) showed a correspondingly high retention of DDT in his body fat after dosing was discontinued. The correlation is an expected one. However, the correlation did not apply during the period of dosing, when three other men who finished the study and one man who failed to finish had higher final DDT values than those in case 24. Therefore, it is not certain that the difference was real, much less that it was associated with race rather than individual difference. There were not enough Negroes in the study to permit testing whether they differed statistically from white people in their ability to excrete DDA. The results in case 26, involving another Negro, prove there is marked overlapping in the ability of the two races to excrete DDA and, therefore, their tendency to retain DDT.

Conclusion

It was shown by Duggan[30] that the average intake of 19-year-old men in the general population during 1964 to 1967 was 0.028 mg per man per day for DDT and 0.063 mg per man per day for total DDT-related compounds. The first of these rates is 1,250 times less than the dosage of p,p'-DDT administered in this study. Even the rate of intake by the general population, which involves the less toxic DDE and DDD as well as DDT, is 555 times less than the dosage of p,p'-DDT we have given. In view of the facts that the storage of DDT is proportional to dosage and no definite clinical or laboratory evidence of injury by DDT was found in this study, these factors indicate a high degree of safety of DDT for the general population at past or current rates of intake.

References

1. Hayes WJ Jr, Durham WF, Cueto C Jr: The effect of known repeated oral doses of chlorophenothane (DDT) in man. *JAMA* 162:890-897, 1956.

2. Walker KC, Goette MB, Batchelor GS: Dichlorodiphenyltrichloroethane and dichlorodiphenyldichloroethylene content of prepared means. *J Agric Food Chem* 2:1034-1037, 1954.

3. Cueto C, Barnes AG, Mattson AM: DDT in humans and animals: Determination of DDA in urine using an ion exchange resin. *J Agric Food Chem* 4:943-945, 1956.

4. Hayes WJ Jr, Dale E, Pirkle CI: *Effect of Known Repeated Oral Doses of DDT,* Working Paper 6.12. Expert Committee on Insecticides (Toxic Hazards to Man), Geneva, World Health Organization, 1961.

5. Durham WF, Armstrong JF, Quinby GE: DDA excretion levels: Studies in persons with different degrees of exposure to DDT. *Arch Environ Health* 11:76-79, 1965.

6. Hayes WJ Jr: Monitoring food and people for pesticide content, in *Scientific Aspects of Pest Control,* publication 1402. Washington, DC, National Academy of Sciences–National Research Council, 1966, pp 314-342.

7. Hayes WJ Jr: Toxicity of pesticides to man: Risks from present levels. *Proc Roy Soc Biol* 167:101-127, 1967.

8. Schneider AJ, Willis MJ: Sources of variation in a standardized and a semi-micro procedure for the spectrophotometric assay of serum glutamic-oxaloacetic transaminase concentration. *Clin Chem*

4:392-408, 1958.

9. Krebs HA, Roughton FJW: Carbonic anhydrase as a tool in studying the mechanism of reaction involving H_2CO_3, CO_2, HCO_3. *Biochem J* 43:550-555, 1948.

10. Yaffe CD, Jones HH, Weiss ES: Industrial noise and hearing loss in a controlled population: First report of findings. *Amer Industr Hyg Assoc J* 19:296-312, 1958.

11. Tentative standard procedure for evaluating the percentage loss of hearing in medicolegal cases, COUNCIL ON PHYSICAL MEDICINE. *JAMA* 133:396-397, 1947.

12. Khairy M: Changes in behaviour associated with a nervous system poison (D.D.T.). *Quart J Exp Physiol* 11(pt 2):84-91, 1959.

13. Rider JA, Hodges JL, Swader J, et al: Plasma and red cell cholinesterase in 800 "healthy" blood donors. *J Lab Clin Med* 50:376-383, 1957.

14. Hayes WJ Jr: Pharmacology and toxicology of DDT, in Müller P (ed): *DDT, The Insecticide Dichlorodiphenyltrichloroethane and Its Significance.* Basel, Switzerland, Birkhäuser Verlag, 1959, vol 2, pp 9-247.

15. Laug EP, Kunze FM, Prickett CS: Occurrence of DDT in human fat and milk. *Arch Industr Hyg* 3:245-246, 1951.

16. Hayes WJ Jr, Quinby GE, Walker KC, et al: Storage of DDT and DDE in people with different degrees of exposure to DDT. *Arch Industr Health* 18:398-406, 1958.

17. Rothe CF, Mattson AM, Nueslein RM, et al: Metabolism of chlorophenothane (DDT): Intestinal lymphatic absorption. *Arch Industr Health* 16:82-86, 1957.

18. Schechter MS, Haller HL: Colorimetric tests for DDT and related compounds. *J Amer Chem Soc* 66:2129-2130, 1944.

19. Dale WE, Copeland MF, Hayes WJ Jr: Chlorinated insecticides in the body fat of people in India. *Bull WHO* 33:471-477, 1965.

20. Dale WE, Quinby GE: Chlorinated insecticides in the body fat of people in the United States. *Science* 142:593-595, 1963.

21. Laws ER Jr, Curley A, Biros FJ: Men with intensive occupational exposure to DDT: A clinical and chemical study. *Arch Environ Health* 15:766-775, 1967.

22. Ortelee MF: Study of men with prolonged intensive occupational exposure to DDT. *Arch Industr Health* 18:433-440, 1958.

23. Booth K, Hancock PET: A study of the total and differential leucocyte counts and haemoglobin levels in a group of normal adults over a period of two years. *Brit J Haemat* 7:9-20, 1961.

24. Osgood EE, Brownlee IE, Osgood MW, et al: Total differential and absolute leukocyte counts and sedimentation rates. *Arch Intern Med* 64:105-120, 1939.

25. Neal PA, von Oettingen WF, Smith WW, et al: Toxicity and potential dangers of aerosols, mists, and dusting powders containing DDT. *Public Health Rep*, suppl 177, pp 1-32, 1944.

26. Torda C, Wolff HG: Effect of convulsant and anticonvulsant agents on the activity of carbonic anhydrase. *J Pharmacol* 95:444-447, 1949.

27. Davenport HW: Carbonic anhydrase in tissues other than blood. *Physiol Rev* 26:560-573, 1946.

28. Robinson J: Dynamics of organochlorine insecticides in vertebrates and ecosystems. *Nature* 215:33-35, 1967.

29. Robinson J: The burden of chlorinated hydrocarbon pesticides in man. *Canad Med Assoc J* 100:180-191, 1969.

30. Duggan RE: Residues in food and feed: Pesticide residue levels in foods in the United States from July 1, 1963, to June 30, 1967. *Pest Monit J* 2:2-46, 1968.

A Case
Of Human
Pesticide Poisoning

BY W. A. GILPIN, M.D.

In 1959 following an aerial spray program in Northwest Detroit by the Michigan State Agricultural Department for Japanese beetle infestations, an awareness of pesticide poisoning became evident. Four patients with mild toxicity consisting of respiratory symptoms were seen. They ranged in age from 12 to 50. Medical literature was sparse on this subject so correspondence with Rachel Carson, Ph.D., the author of "Silent Spring," was initiated. The four cases were recorded in her book and an awareness of a new medical problem ensued.

Since World War II, the chemical industry has produced ever increasing quantities of pesticides in great varieties and strength. It is the so-called hydrocarbons (DDT, Aldrin, Dieldrin, Entrin, Chlordane, Lindane, Heptachlor, Toxaphene) that have become harmful to wild life as well as humans. D.D.T., the commonest and the most widespread, is the hydrocarbon this article will pursue.

United States chemical companies began to produce DDT in high volumes in the 1940's.

1948-1949	38,000,000 lbs.
1951-1956	110,000,000 lbs.
1956-1961	152,000,000 lbs.
1961-1966	150,000,000 lbs.
1967	103,000,000 lbs.

There is now a downtrend in their production. It has been estimated that 46,000,000 lbs. of pesticides are used annually in the United States covering 90 million acres, with one acre out of 10 revealing DDT in the soil. This is a $20 million per year business for the chemical industry.

DDT HAS thoroughly permeated our environment. It has been by far the most widely used pesticide and is toxic to a broad spectrum of animals including man. It is found in water, the air, soil and even in the gulls in the Antarctic. It is found in adipose tissue of man world-wide. Evidence portrays that it is spread by wind and water much as radioactive fall-out. Migrating fish and birds transport it thousands of miles.

DDT has a low solubility in water, (upper limit one part per billion) but as algae and other organisms in the water absorb the substance in fats, it becomes highly soluble. DDT is extremely stable in the environment with a slow breakdown. DDT has a long-life in trees and soil in the range of tens of years.

Ecologists have become aware that pesticide residuals have now accumulated to levels that are catastrophic for certain animal populations, particularly carnivorous birds. The pesticides also destroy predators and competitors that normally tend to limit proliferation of the pest. Pests tend to develop new strains that are resistant to DDT Our chemical onslaught is resulting in the extinction of certain species of wild life, notably carnivorous and scavenging birds, e.g., Hawks, Eagles, and Osprey. Destruction of species balance, reproduction, and behavioral alterations in wild life have been proven. Added pollution to the environment is a fact.

Federal seizure of ten tons of Coho salmon in February 1969, from Lake Michigan verified the fact that DDT levels were over presumed safe limits. The United States Bureau of Commercial Fisheries warned that DDT levels would interfere with reproduction of lake trout, salmon, and other species. As a result, the Michigan Department of Agriculture banned the use of DDT in April 1969; but, chemical DDT could be sold and used in Michigan if in transit from the manufacturer by June 27, 1969.

New research into tolerance levels in humans will be studied. The effect of killing non-target organisms, accumulation in the food chains, lowered reproductive potential, resistance to DDT sprays, synergistic effects, chemical migration, accumulation in the ecosystem, and delayed responses are to be studied.

CONSERVATION GROUPS and many scientific groups are well aware that DDT affects wild life, and will eventually cause side effects in humans. It is estimated that four lbs. of pesticides are used for every man, woman, and child in America each year. It becomes an enormous potential problem from a health point of view.

At a Michigan State Medical Society meeting in the fall of 1965, Herman Kravill, Chief of the Pesticide Control Committee for the United States Department of Health, indicated that there were 60,000 pesticides on the market. Over 100 of these were toxic. Twenty percent of all pesticides were used in the state of California. Of all the pesticides used in the United States at that time, the majority were largely hydrocarbons, or organic phosphorous compounds. He went on to establish the fact that DDT was found in 80 parts per million in trout. As a result, many died and thousands developed cancer-like tumors.

He went on to say that pesticides were appearing in our meats, seafoods, eggs, milk, fruits and vegetables. Thirteen p.p.m. of DDT were found in human milk, whereas cow's milk had three quarters less DDT. In humans country-wide at that time, there were 16.52 p.p.m. of DDT in adipose tissue, 1.44 p.p.m. in the liver, 1.5 p.p.m. in the kidney, 1.50 p.p.m. in the brain, .45 p.p.m. in the gonads. This established the fact that humans did carry DDT.

A reproductive failure in certain birds was shown then, and has been proven since in Pheasants, Mallards, Eagles, Osprey and Hawks. The effects of DDT have no antidote and it is now thought the average level of DDT in humans, in p.p.m., is 12.90 percent largely derived from food. It is now known that DDT is passed through the placenta to the fetus.

LABORATORY EXPERIMENTS with animals indicates DDT attacks the central nervous system, upsets body chemistry, distorts cells, accelerates gene mutation, reduces drug effectiveness and affects calcium absorption by the bones. Recent Hungarian experiments with mice emphasized DDT's carcinogenic possibilities and they have banned its use. Carcinogenic properties were confirmed by the National Cancer Institute's interim report on the long-term toxic effects of 130 chemical compounds. Specifically it tagged 11 common pesticides (including DDT) and cast strong suspicion on 19 other chemicals as sources of tumors in mice. It also stated that it is reasonable to conclude, the great mapority of the DDT-induced tumors have malignant potentiality.

Kenneth P. Du Boss, professor of pharmacology in the Pritzker School of Medicine of the University of Chicago discovered that insecticides including DDT can stimulate the liver to produce abnormally high levels of certain enzymes that cause increased detoxification of drugs that may render the drug therapeutically ineffective at normal dosage levels. Tests in the toxicity laboratory conclusively demonstrated that frequently used sedatives as barbiturates are counter-acted by DDT.

It is of interest to note that 70 workers exposed to DDT for ten years in a Soviet Study did show increased acid pepsin secretion in the stomach; in the liver there was pronounced disturbance of protein and sugar metabolism with pigmentation, detoxication and secretion functions. The United States Department of Public Health Study recently noted that fairly high doses of DDT in the diets of rats formed liver malfunctions, but monkeys in a similar study revealed no ill effects.

It has been demonstrated that DDT can cause adaptation of enzyme systems although the general biological significance is unknown. DDT in rats, it is noted, is an "enzyme inducer," and not only breaks down drugs but a number of different substrates, including estrogen. In birds, estrogen mediates calcium metabolism causing the birds to accumulate calcium in the hollow parts of the skeleton for later transfer by way of the blood stream to the oviduct, where the egg shell is formed. Thus, it was hypothesized that by breakdown of estrogen and upsetting calcium physiology, DDT was responsible for avian reproduction failures.

Also in the past few years it has been discovered that the DDT induces certain enzymes in the liver that decompose steroid sex hormones creating a hormone imbalance. DDT levels in the human environment caused a similar breakdown in several laboratory animals, including rats, whose sex hormones, estradiol and testosterone, are thought to be identical to those of man. Human cells (Chang liver and He La cells) have been exposed in cell culture to DDT. DDT proved toxic in the cells, and induced progressive morphological changes leading to cell destruction (Gablips and Friedman, Massachusetts Institute of Technology).

DDT has been used as a wettable powder of 25% to 75%, an emulsifiable concentrate of 50% or less, in aerosol bombs, and as a dusting agent in various concentrations. Emulsions and wettable powder suspensions of five percent have been commonly used in spray operations. Dusting applications have contained 10 percent of DDT and many household insecticide sprays have contained five percent solution of DDT in purified kerosene. Absorption is through the alveoli and intestinal tract and through the skin as a solution of fats and oils.

DDT CAUSING illness in man is unknown. Cases of DDT 648 p.p.m. in adipose tissue of five years in man have caused no injury. However, the tolerance is extremely variable from person to person. Symptoms usually begin in 30 minutes to 3 hours with malaise, headaches, fatigue, parathesias of the tongue, lip and face. With large exposures, parathesias of the extremities, extreme apprehension, disturbances of equilibrium, dizziness and confusion with tremor, and convulsions develop in serious cases. Pupils become dilated and there is increased sensitivity to touch and pain in areas of parathesias. Loss of vibration sensation can occur in the fingers and toes. Coordination tests may show poor results. Mild cases develop increased pulse rates while more severe exposures result in slow irregular pulses. Transient jaundice has been reported rarely.

Laboratory findings are usually negative. Urinalysis for DDA (a metabolite of DDT) are positive where concentrations of DDT have been severe. Blood for analysis may be helpful but only the United States Public Health Toxicology Laboratory in Atlanta, Georgia is equipped to run these tests. Cholinesterase levels before DDT exposure and after may be helpful. Thus laboratory tests in humans are available but highly specialized to determine results.

In 1963, with a naturalist, Dr. Walter Nickell from the Cranbrook Institute of Science, we began warning the Forestry Department of the City of Detroit, of the hazards of using DDT continually in the spray programs for Dutch Elm Disease. This operation had progressed since 1957 twice yearly involving 400,000 Elm trees, using a 25% emulsion of DDT in water forced out with spray mist machines. Several meetings with the City Council and the Forestry Department resulted in little progress until last year when a sanitation program (cutting all diseased trees) and a chemical substitute, Methoxichlor, began to be used. Breakdown of this latter chemical was far more rapid. Sanitation alone is still being refused by the city even though this is the best treatment in the long run from a total ecological system approach.

A Mrs. J. A., a 56-year-old school teacher, on April 12, 1962 came in direct contact with DDT sprayed by the City of Detroit for Dutch Elm Disease. The wind blew the DDT emulsion over the front of her house and surrounding shrubs, covering her windows as well. While immediately trying to clean the DDT off her windows she had her right hand pricked multiple times by brushing by a Phitzer bush while moving the step ladder.

She washed windows for one hour and then developed dizziness, ringing in her ears, and began "shaking all over." A cough developed along with "blood shot" eyes with burning. She became nauseated and three hours later developed numbness on the right side of her face, right arm, back and abdominal area with marked twitchings in the areas. Her fingers tingled and she developed small blisters at her finger tips with marked swelling of her right hand, double the normal size. She became extremely stimulated and apprehensive and had a sleepless night.

THESE SYMPTOMS continued for several days accompanied by partial paralysis of her hand with inability to button her clothes, iron, wring out wash or open doors. Marked fatigue developed and despite Epsom Salt soaks her hand continued to remain severely "numb and swollen."

In May 1962 she consulted her family physician who prescribed tetracyclines and cortisone with improvement of edema in one week. Her other symptoms continued. She then consulted a dermatologist at Henry Ford Hospital who noted a slight residual swelling of her hand and a faint erythema. She appeared very nervous. No other signs were noted. CBC was normal except for a normacytic anemia. Urinalysis was negative, and VDRL was negative. Antihistamines with mild sedation were prescribed. The dermatologist felt she had a lymphedema associated with a contactant to be found in a menopausal nervous individual. Although her pulse was normal, her blood pressure had increased in nine days from 130/85 to 140/100. He then added Theelin. Little change occurred in her hand except she felt the Theelin had helped her anxiety state. She was referred to the Arthritic Clinic but did not return.

After she read a newspaper article about DDT toxicity, she was referred by Cranbrook Institute of Science and was seen on February 21, 1963 for the first time. Her subjective complaints were extreme nervousness, fatigue, twitching of the right facial area, right eye, right hand, right arm, and intermittent twitching of her back and abdominal muscles. Her hand, she complained, was partially paralyzed, sore, stiff, swollen each morning, and she was unable to close it. Her grip was poor. She still had occasional dizzy spells, headaches and cramps in her legs at night.

EXAMINATION REVEALED a fairly well developed, well nourished middle aged white female in no acute distress, but extremely nervous. Blood pressure was 150/102. Positive findings revealed slight stiffness of the fingers of her right hand and slight tenderness of the entire right hand to palpation. Her grip was 50% less than in her left hand. The entire function of her right hand was 30% of normal. Neurological examination was negative except for generalized hyperactive reflexes with marked hyperactivity of the right upper extremity. There was some loss of muscle tone. Sensation to touch and temperature changes were normal. Co-ordination of the finger to nose test was only 50% of normal. Temperature, pulse, and respirations were normal. Petite muscular twitching of her right facial, right eyelids and right arm muscula-ture was noted.

The rest of the complete examination was nega-tive except for a slight anemia 3,300,000 red blood cells with a hemoglobin of 11.5 grams. White blood count was normal as was urinalysis. BSP was borderline. Cephalin Flocculation was 3+. Chest X-Ray and ECG was normal. Urine was sent to the Communicable Disease Center Labora-tory in Atlanta, Georgia, for DDT and returned April 25, 1963 revealing a negative DDA (DDT metabolite) result. This was a microcoulometric gas chromatography test using a gaseous color-meter process.

Impression at this time despite DDA test was that she had DDT poisoning resulting in:

1. 30% paralysis and decreased muscle tone of right hand
2. Residual Central Nervous System damage with hyperactive reflexes probably from DDT poisoning
3. Mild Secondary Anemia
4. Subclinical Liver Damage (hepatic) on basis of borderline liver function

From two other cases I felt these impressions were accurate. Therapy consisted of moderate sedation, iron, and B complex in large doses over a period of four months and resulted in an im-provement of 50% of the paralysis of her right hand, 25% of her motor coordination, but resid-ual CNS damage still revealed hyperactive reflexes, anxiety and stimulation. Liver function reverted to almost normal findings. Her cephalin floccula-tion was 1+ with a normal BSP. However, her blood pressure remained 150/100. Her anemia improved to 3,600,000 and hemoglobin 11.8 grams. I felt these present residuals would remain. This case was tried in Detroit in June 1968 and won after a week of testimony by the chemical company involved with the City of Detroit and the Plaintiff. An appeal by the City and by the chemical company in the State Appeals Court was dismissed.

THIS CASE illustrates what can happen with the continued use of DDT without its ban. Even though DDT and chemically related insecticides have made enormous contributions to public health and agriculture, it is building up resistant insect populations in many areas and upsetting ecosystems. Substitute compounds can be used. DDT should be extinct.

Interior Secretary Walter Hickel has announced that the Federal Water Pollution Control Administration awarded a contract to the Aerojet-General Corporation to develop a self-destruct catalyst for DDT. The Secretary hopes to prevent future deposits of DDT from endangering the environment and to degrade the toxicity of pesticides already in the soil being used for human consumption.

A bill to prohibit the manufacturing or sale of DDT throughout the nation is now in Congress. With environmental pollution we must be protected or man will destroy himself. We as physicians must become aware of pesticide toxicity in man and attempt to support research that is constantly widening our knowledge into recognition of its human effect.

BIBLIOGRAPHY

1. *Clinical Handbook on Economic Poisons,* U.S. Dept. of Health, Education and Welfare.
2. Gibbons, Harry, M.D., *Some Medical Aspects of Pesticide Poisoning 11 A-23,* Oct. 23-24, 1967.
3. Michigan Department of Conservation Bulletin 1/18/68 and 2/15/68.
4. Pesticides Monitoring Journal, March 1968, Vol. 1, No. 4.
   Pesticides Monitoring Journal, June 1968, Vol. 2, No. 1.
   Pesticides Monitoring Journal, September 1968, Vol.

2, No. 2.

Pesticides Monitoring Journal, December 1968, Vol. 2, No. 3.

Pesticides Monitoring Journal, March 1969, Vol. 2, No. 4.

Pesticides Monitoring Journal, June 1969, Vol. 3, No. 1.

5. *Storage of DDT and DDE in People with Different Degrees of Exposure to DDT*, A.M.A. Archives of Industrial Health, Vol. 18, Nov. 1958.

6. West, Irma, M.D., *Pesticides as a Public Health Problem in California*, Oct. 1963.

7. West, Irma, M.D., *Occupational Experiences with Pesticides in California*, 1964.

8. West, Irma, M.D., *Pesticides as Contaminants in California*, 1964.

Fact and Fancy in Nutrition and Food Science[1]

THOMAS H. JUKES

THE THEME OF this convention, I was told, is "The Dietitian in the Age of Aquarius." This seems a highly appropriate title, reflecting as it does the current preoccupation with the signs of the Zodiac—a turning back of the clock to the age of mysticism and superstition. I recently saw a cartoon which showed two loan desks in a public library, one for astronomy and the other for astrology. The astronomy desk was deserted, but many readers were crowded around the other desk. I am reasonably convinced that we stand on the threshold of a new Dark Age, which we shall enter unless, perhaps, enough of us can keep the light of scientific knowledge burning—for, without science and technology, our food supply will soon dwindle and few things cause human beings to break down as rapidly as an insufficiency of food.

Much publicity has been given to the announcement that a majority of students preferred organically-grown foods, when offered a choice in a campus restaurant on the Santa Cruz campus of the University of California. Either no one has told them, or they don't believe, that there is no such thing as organic plant nutrition. Plants utilize only inorganic forms of plant food. Compost and manure are broken down by bacteria to components, such as nitrate, potassium ion, and phosphate, before they are assimilated. Hydroponics, in which plants are grown inorganically without soil, leads

[1]Presented at the joint meeting of the California and Nevada Dietetic Associations in Las Vegas on April 19, 1971.

to the production of vegetables and fruits with the same protein, carbohydrate, vitamin, and mineral content as when the same strains of plants are grown in the ground with lots of manure. An important difference is that organically grown vegetables are more likely to carry Salmonella, which leads to a common form of food poisoning. Salmonella has recently been reported to be present in vegetables in Holland, grown with the organic benefits of sewage effluent (1).

The book, *Silent Spring* (2), which contained pleasant statements, such as "We are in little better position than the guests of the Borgias," is used in many public and high schools as an authoritative text. Many imitators have followed the same approach in fomenting alarm against agricultural chemicals, and, as a result, the public is questioning the quality of our abundant and nutritious food supply, the safest, cheapest, most diverse and best in the history of the world. I say cheapest in terms of hours of work necessary for the average person to exchange for his daily food supply when purchased in the open market. Only within the past forty years or so has it become possible to buy numerous delicious fruits, vegetables, meats and dairy products free from contamination, fresh, canned or frozen, at any time of the year. Remember that we are only two generations away from the days when food was frequently polluted with cockroaches, rat filth, insect parts, tuberculosis bacteria, meal-worms, weevils, and molds. The fungus, *Phytophthora infestans*, that destroyed the potato crop in Ireland and caused the famine, is still around. A sample of what could happen on a large scale was seen last year when another blight swept through many corn fields in the eastern United States. As a result of scientific agriculture, in 1971 a broiler chicken costs less in *actual, inflated* cents per pound than it did in cents per pound in the depth of the Depression.

Unless we are vigilant, however, all this will vanish, dissipated by a destructive and vociferous few who have grown up to think that food comes right out of the floor of the supermarket. Two years ago, I said in a talk in Seattle: "Perhaps man cannot understand how precious food is unless he has to toil to produce it himself, fighting against drought, blight, and insects for the privilege of getting something to eat. Can we learn without repeating such bitter experiences?" I believe that dietitians have a great responsibility to instruct and lead the new urban generation in the facts of nutrition and food science.

Some Toxicologic Principles

Let me emphasize certain fundamental points regarding measurement of the effects of chemicals. First, tests carried out by injecting a chemical are usually meaningless in terms of residues present in foods. Injecting sodium glutamate, or DDT, or distilled water into baby rats will injure or kill them, especially if the quantity is large. Therefore, the first rule is that the route of administration must be by mouth.

Next comes the quantity used. It is astonishing to see, repeatedly, newspaper articles with scare headlines describing the toxic effects of some enormous dose of a chemical such as 2,4,5-T, inferring that this incriminates tiny amounts in food. Several essential nutrients or metabolites are recognized poisons at levels that do not greatly exceed the daily requirement or the amount naturally in the body. Examples are copper, iodine, selenium (which is a carcinogen), table salt, hydrochloric acid, and hydroxyl ion.

The next point, which is frequently raised when it is pointed out that the dosage is small, is the cumulative effect. Alarmists often say that traces of any residue will be retained by the body and will build up to dangerous levels. Actually no completely cumulative effect exists. There is always some excretion, and the total amount stored represents the balance between intake and excretion. Let me quote Professor Wayland Hayes, Jr. (3):

> Compounds which are absorbed tend to reach a steady state of storage and for each compound the storage at equilibrium corresponds to dosage. People not trained in the biological sciences are usually unfamiliar with these facts and suppose that compounds are of two distinct sorts, those that are stored and those that are not. They suppose that compounds that are stored at all continue to accumulate indefinitely with no tendency to reach a steady state in which the amount lost each day is equal to the amount absorbed. It may seem odd to mention this folklore . . . but the views of the public are a most important factor, which must be taken into account in any long-term effort to achieve general public acceptance of a particular usage or course of action.

The balance or reservoir of a chemical in the body will depend on the species of animal, the rate of metabolism of the substance, and its chemical properties. One compound that is most frequently under attack as being stored in the body from food residues is DDT. DDT is not cumulative. It is stored in the fat and is steadily broken down and excreted in the urine. If the daily intake goes down, the level in the fat drops. There is a straight line relationship which has been worked out in several careful experiments, as shown by Hayes (3). The average amount of DDT in the fat of the general American population is progressively dropping, year by year because the amount used in agriculture has been reduced (Table 1).

The next point is one that I should probably have mentioned first: How reliable is the test method? There is a delicate analytical procedure called gas liquid chromatography with electron capture. Sometimes I wonder whether this method, in the hands of inexpert people, has done more harm than good. There has been a great hue and cry over alleged traces of DDT in Antarctic penguins, amounts of the order of 1 or 2 parts per billion. I have not yet been convinced of the validity of the results. A few months ago, at the University of Wisconsin, some soil samples that had been sealed since 1910 were tested for synthetic organochlorine pesticides by the latest and most delicate gas chromatographic procedure. Several pesticides were detected in thirty-two of the thirty-four samples (5). The only flaw was that these pesticides not only were not used in 1910, they didn't even exist until after 1940! Another complication is that residues of a class of modern compounds called polychlorinated biphenyls (PCB's) interfere with the DDT test (5). The PCB's are used in water-proofing compounds, asphalt, waxes, synthetic adhesives, hydraulic fluids, electrical apparatus, and generally in plastics. They are widely distributed in the fat of wildlife species, in which they have originated as industrial wastes taken up by aquatic species. They overlap in the test with DDT and its metabolic breakdown products, DDD and DDE. PCB's are sufficiently toxic to kill fish in hatcheries (5). To sum up, PCB's are not used as pesticides, but they interfere with pesticide residue analysis, and they are toxic.

Thus, I don't believe the stories of "newspaper scientists" about pesticide residues until they have been published in the scientific literature, scrutinized, and reliably confirmed.

Mercury

This brings us to the current concern over mercury: Has it always been there anyway? In this case: Yes.

Sources of mercury may be either industrial and technologic or natural (6). The industrial sources include coal smoke, chlor-alkali plants which use mercury electrodes, plastic manufacture, fungicides (such as methyl mercury dicyanodiamide, used as a seed dressing), slimicides (used in the paper industry), and electrical equipment. About 200 gm. mercury may be used per ton of chlorine manufactured, and it has been estimated to add up to a total of 300,000 lb. mercury annually in Canada and 500,000 in the United States. A large amount of this got into the Great Lakes and information on mercury in fish was published in Canada after eighteen months of government investigation. It was disclosed that unacceptably high levels of mercury were present in many fish caught in inland waters where they had ingested methyl mercury formed by bacteria in bottom muds. The industrial use of mercury in Sweden appears to have led to decreases in bird populations, together with paralytic symptoms in the birds, which, of course, were blamed on DDT. However, the most tragic incident of mercury pollution was in Japan, where more than fifty people, including babies, died in the 1950's from "Minamata disease," caused by eating fish and shellfish contaminated with mercury in the effluent of a chemical plant. Some of the shellfish sampled from 1956 to 1958 contained 9 to 24 p.p.m. (wet basis); a level of 5 to 6 p.p.m. in the diet is regarded as potentially lethal.

The second source of mercury is natural. Interest in this was triggered by industrial contamination, following which some samples of canned tuna were found to contain about 1 p.p.m. mercury. This set off a shock wave of publicity, and 12.5 million cans of tuna were removed from the market. It was then reported that mercury concentrations in tuna caught ninety years ago are about the same as those in recently caught fish (6). It was also found that swordfish commonly contained 1 or 2 p.p.m. mercury, even when they were caught in ocean waters hundreds of miles from possible industrial contamination.

The presence of mercury in sea water has been known and measured ever since 1777. All the metals known in the earth's crust, including gold, are present in sea water. Some marine organisms concentrate them from sea water; for example, shrimp concentrate arsenic to a level "higher than the government allows." Seaweed picks up iodine. And, apparently, swordfish and tuna frequently concentrate natural mercury to a point higher than the tolerance set by the Food and Drug Administration. The residues are higher in older fish. The following statement was in the *Marine Pollution Bulletin* for January 1971 (7):

> If one considers that there is a natural occurrence of mercury in the sea of between 0.03 and 0.3 parts per billion and that tuna fish has a higher respiration rate than most other edible fish, swimming continuously and passing large quantities of water over its gill surfaces, and that it can accumulate metals in its tissues at concentrations thousands of times greater than those in the water, it could be that the phenomenon we are observing is caused not by levels of mercury increased by pollution, but by the unusual susceptibility of the tuna fish to the naturally occurring mercury.

Such a situation is quite difficult for regulatory authorities to accept mentally. For example, FDA Commissioner Charles C. Edwards was quoted (8) as follows:

Q. What about the mercury content in swordfish?
A. We're wrestling with that. We're not quite as far along as with tuna, but I think before too long we'll be able to say that all swordfish left on the market is clean.
Q. Is there some on the market now that isn't?
A. I suspect there is. Again, we're working with the industry. Some 85 to 90 per cent of the swordfish that we've been testing has been running over our guideline.
Q. How much over?
A. Quite a little more than tuna. Some samples have been running over 1 part per million. But swordfish is a somewhat different problem than tuna because the consumption in the U. S. is considerably less.

This remarkable piece of semantic gymnastics should be examined in the light of the facts that mercury is toxic, regardless of its origin, that "natural" mercury in swordfish is commonly above the "safe" level set by the FDA and that this level (0.5 p.p.m.) is ten times as high as that proposed to the World Health Organization (0.05 p.p.m.), which would effectively remove most fish as an article of human diet. Deposits of mercury ores, such as cinnabar, are present in the coastal and foothill ranges of northern California, where mercury is mined, and inevitably mercury gets into the streams. On May 2, 1971, for example, the fish in Almaden reservoir, California, were found to contain up to 4 or 5 p.p.m. mercury. The pollution was believed to result from "numerous ancient mercury mines in the area."

Is the FDA tolerance for mercury too strict? It is one

tenth of the potentially lethal level. Compare the
mercury tolerance with the DDT level (0.05 p.p.m.)
permitted in milk and note that levels of DDT thousands of times as high as this have been consumed for
prolonged periods without any evidence of harm (Table 1).

TABLE 1   **Dosage response of DDT in man***

DOSAGE	REMARKS
mg./kg./day	
?†‡	fatal
16-286†	prompt vomiting at higher doses (all poi-soned, convulsions in some)
10†	moderate poisoning in some
6†	moderate poisoning in one man
0.5	tolerated by volunteers for 21 months
0.5	tolerated by workers for 6.5 yr.
0.25	tolerated by workers for 19 yr.
0.004	dosage of Indians, 1964; combined intake from living in sprayed houses and from food
0.0025	dosage of general population of U.S., 1953-1954
0.0004	current dosage of general population of U.S.

*From WHO statement, January 22, 1971 (4).
†One dose only.
‡Dosage unknown.

The reason for this incongruity is, of course, that
condemning nature is unpopular. The dogma of the en-
vironmentalists seems to be that "only man is vile."
Puffer fish, blowfish, loco weed, and rattlesnake bites
are all both "natural" and toxic. Mercury poisoning is
especially unpleasant because it causes brain damage
and affects unborn babies.

What are we going to do about it? We will have to
establish a tolerance for mercury and make a mental ad-
justment to it, for mercury, like all the elements, is
naturally present in all living organisms, including fish.
It is most interesting that mercury residues were first
shown to come from industrial pollution; then, as a result of this discovery, levels of mercury in fish, arising
from natural sources, were shown to reach potentially
toxic proportions.

The "Delaney Clause"

Our next point is the celebrated Delaney anti-cancer
clause in the federal Food Drug and Cosmetic Act. The
effect of this is to require the removal from interstate
commerce of any food which contains analytically detectable amounts of a food additive shown to be capa-

ble of inducing cancer in experimental animals (9).

This law is impossible to enforce and impossible to repeal. Rivers of ink have been spilled on it. Many naturally occurring food ingredients are known to be carcinogenic (10) and at least one essential nutrient, selenium, is a carcinogen (11). Iodine *deficiency* will produce tumors in experimental animals. All the steroid sex hormones are carcinogenic in animals, and so is cholesterol. The barbecuing of meat produces identifiable carcinogens.

The Delaney clause does not permit tolerances; thus any detectable amount, even one molecule, comes within the meaning of the Act. It is sometimes argued that there is no threshold level for a stimulus that produces cancer, but this concept cannot be true for the tumors produced by iodine deficiency. Nor can it be true for selenium. Furthermore, the dose-response curves plotted against time sometimes show that at low levels of a carcinogen, the predictable occurrence of cancer would take more than a lifetime to be reached.

Similar problems exist with respect to mutagenic and teratogenic substances because these are always present in foods, and mutations are a normal process of all life. Evolution cannot take place without the constant occurrence of mutations.

DDT and Malaria — Present Status

DDT is specifically needed to protect millions of people in tropical countries from death by malaria. This has repeatedly been made plain by the World Health Organization, for example (12):

> The withdrawal of DDT would mean the interruption of most malaria programs throughout the world . . . DDT used as a residual spray of the interior surfaces of houses . . . led to the idea of nation-wide malaria control campaigns including the whole of the rural areas of a country. The success of these campaigns resulted in the concept of malaria eradication which was adopted . . . for the world by the World Health Assembly in May, 1955.
>
> Since then DDT has been the main weapon in the world-wide malaria eradication program. Research has continued for the development of other methods of attack against malaria and for the development of alternative insecticides. To date, there is no insecticide that could effectively replace DDT which would permit the continuation of the eradication program or maintain the conquests made so far.
>
> The withdrawal of DDT will represent a regression to a malaria situation similar to that in 1945. The reestablishment of malaria endemicity would be probably attained following a period of large scale outbreaks and

epidemics which would be accompanied by high morbidity and mortality due to loss of immunity by populations previously protected by eradication programs.

This prediction has been fulfilled in Ceylon (4). To continue (12):

> Toxicological observations of spraymen working for a number of years in malaria eradication and even in formulation plants, has not revealed toxic manifestations in them. Neither has there been any evidence of toxicologic manifestations in people residing in houses that have been repeatedly sprayed at six-month intervals.
>
> We therefore believe that a great harm will result from the unqualified withdrawal of DDT. We feel that selective use of DDT is justified and warranted.

This is what the argument is all about. If the manufacture and export of DDT are banned in the United States, the world-wide anti-malarial program will collapse. Most of the DDT manufactured domestically is for this program. Furthermore, a ban in the United States would lead to prejudice against the use of DDT elsewhere.

The substitute insecticides are more expensive and more profitable than DDT. These substitutes can be used, with varying degrees of lower efficiency, against the agricultural pests controlled by DDT. But, there is no effective substitute for DDT in the world campaign against malaria. The other compounds either decompose rapidly, produce resistance too fast, or they are too poisonous to people.

Malaria is caused by a microscopic parasite of the genus *Plasmodium*. These parasites spend part of their life-cycle in mosquitoes, but the cycle is not complete without going through several stages in man, where they reach maturity in the red blood cells and reproduce, in enormous numbers, into a form called *merozoites*. These change into the sexual stage which enters the body of a blood-sucking mosquito. Other stages of the life-cycle then take place, and the parasite reaches the salivary gland of the mosquito, from which it is inoculated into the next victim to be bitten. The cycle then continues from man to mosquito and mosquito to man. The principal method for breaking this pernicious chain is to kill mosquitoes with DDT by spraying the interior walls of human dwellings.

Public health authorities in the tropics commonly use a figure of 1 per cent per year to estimate the mortality from malaria: thus the 75 million cases in India were calculated to be responsible for 750,000 deaths annually (33). The survivors in many cases are

116

severely debilitated and unable to work. The effects aimed for in the mosquito control program are based on the following conditions: The mosquitoes rest on the walls by day and attack sleeping people at night. The DDT on the walls kills the mosquitoes, and for this purpose, an insecticide must be *persistent* because it is not possible for spray teams to go into the same house frequently.

As a result of international cooperation, WHO has had a world-wide malaria eradication program so that (14):

> Today more than 960 million people who a few years ago were subject to malaria endemicity are now free of malaria; another 288 million live in areas where the disease is being vigorously attacked and transmission is coming to an end. Because much of Africa remains highly malarious and because about 288 million people live in malarious areas not yet subject to eradication measures, it is logical that the United States should maintain an active interest in this disease.

These estimates indicate that the U. S. contribution to saving lives from malaria has paid off quite well in terms of human welfare. As I said recently, however, "Some Americans, by demanding a ban on DDT, are reversing the traditional role of their country in relieving the sufferings of others" (15).

DDT in Human Milk

One of the organizations leading the fight on DDT is the Environmental Defense Fund (EDF). A recent newspaper article on the EDF stated:

> The turning point came when Cameron decided to spend $5,000 of the organization's total remaining assets of $23,000 on an advertisement in the New York *Times* on Sunday, March 29, headlined 'Is Mother's Milk Fit for Human Consumption?' It referred to the amounts of DDT in the human body.
> ·The ad appealed for members, starting at $10 for a basic membership. It produced $7,000, a profit, and the EDF turned to a direct mail campaign and now has 10,000 members, a stable financial base and a chance at major foundation support.

This is most interesting. The EDF appealed to the public on the basis of the DDT content of human milk. As a means of arousing alarm concerning DDT, the EDF and the National Audubon Society have both stated that DDT causes cancer. The implication that DDT in breast milk may cause cancer in babies is superlatively sensational copy. The following lurid passage appeared in *Purple Martin Capital News*, July 29,

1970, as a quotation from an article by Ed Chaney, Information Director of National Wildlife Federation, in *Conservation News*, June 15, 1970:

> A five-day-old human being lies asleep in the other room. His name is Eric. His tiny, wiggly, red body contains DDT passed on to him from his mother's placenta. And every time he sucks the swollen breasts, he gets more DDT than is allowed in cow's milk at the supermarket. Be objective? Forget it. Objective is for fence posts. How can you be objective in the face of a global insanity that is DDT? In the face of abdicated responsibility by the men the public pays to protect its interests? Are the anarchists right? Are ashes the only fertile seed bed for growing new responsiveness to the public interest? Picture a swarm of angry citizens bathed in the light of flames engulfing the Agriculture Department.

It is distressing that an official of a large organization should discard objectivity and propose anarchy in its stead. It is also distressing to read such absolute rubbish when DDT has saved the lives of hundreds of thousands of babies who otherwise would have died from one of the most lethal of diseases, infantile malaria, which Sir Macfarlane Burnet in 1953 called "the main agent of infantile mortality in the tropics." In the years following 1953, this disease was stopped in many countries by DDT.

Let us examine the factual and scientific background for the propaganda campaign regarding DDT in human milk. Improvements in technology have made it possible to detect fantastically small quantities of DDT. But, such extremely delicate tests can easily give "false positive" readings because of accidental contamination of the equipment or lack of expertise by the tester. Next, cow's milk has occupied an unusual position among foods with respect to regulations. "Zero tolerance" has been the policy with respect to additives to milk, except for vitamin D. The improvements in testing procedures made it necessary to re-examine the definition of zero, since every chemist knows that zero content, in molecular terms, does not exist. More than ten years ago, it was evident that the entire stocks of canned milk in the United States gave positive tests for DDT. It was, therefore, necessary to face facts, and two choices were available: to ban cow's milk from interstate commerce or to set a tolerance.

The latter choice was taken, and the tolerance set was 0.05 p.p.m. This was a far lower level than the 7 p.p.m. which was permitted for most agricultural products. A rule-of-thumb for tolerance levels is 1 per cent of the toxic dose which is lethal to 50 per cent of a group of experimental animals. Obviously, if 7 p.p.m. had been estimated to be non-injurious, a tolerance of 0.05 p.p.m. provided an unusually large margin of safety. The low tolerance was possible primarily because cows effectively metabolize and break down DDT, and also because great attention was paid to avoiding the use of DDT on crops, such as alfalfa, which are consumed by dairy cattle. In contrast, human beings are less efficient than cows in metabolizing DDT, and they do not eat hay. There is a straight-line relationship between DDT intake and DDT level in body fat.

The DDT in human beings enters the fat of breast milk. This was noted in 1950 by Laug and co-workers (16), who found an average concentration of DDT of 0.13 p.p.m. in thirty-two samples taken in Washington, D.C., with a range from undetectable ("zero") to 0.77 p.p.m. Several similar reports have since appeared.

The level of DDT in human milk is about twice as high as the FDA tolerance allowed in cow's milk. This statement needs an explanation of its background and meaning in terms of toxicology. Using the unexplained statement to alarm the public, particularly nursing mothers, is a scientifically irresponsible act.

The World Health Organization and the Food and Agricultural Organization of the United Nations set a permissible rate of intake of 0.01 mg. DDT per kilogram body weight for breast-fed infants. The DDT intake of breast-fed babies in the United States may be higher than this; estimates range from 0.014 to 0.02 mg. per kilogram per day at birth, if the infant consumes 600 ml. (about 1½ pt.) of breast milk daily. As the infant grows, the intake of milk on a per-kilogram basis decreases because food intake per unit of body weight lessens when the size of an animal increases and because breast-fed infants usually receive supplementary feeding with other foods.

The "permissible rate" set by the WHO-FAO, according to the chairman of the meeting that established the value, is highly conservative, and he points out (13):

> It offers a safety factor of about 25 compared with what workers in a DDT manuacturing plant have tolerated for 19 years without any detectable clinical effect (see Laws et al., Arch. Environ. Health 15:766-775, 1967). The safety factor of the WHO-FAO permissible rate is 150 compared to the dosage of DDT given daily for 6 months to a patient with congenital unconjugated jaundice with-

119

out producing any side effects (Thompson *et al.*, Lancet II [7610]:2-6, July 5, 1969.)

Infants are more susceptible than adults to some compounds, but the difference is seldom great—usually about 2 to 3 times. In a study of 49 different compounds, newborn rats were found to vary from 5 times less susceptible to 10 times more susceptible than adults. Although there is no information on the relative susceptibility of human infants and adults to DDT, it was shown by Lu *et al.* (Food and Cosmetic Toxic. 3: 591-596, 1965) that weanling rats are slightly more resistant than adult rats to this compound, and that preweanling rats are more than twice as resistant and newborn rats are over 20 times more resistant than adults.

Evidently it is possible for breast-fed infants to obtain DDT from the milk at a level up to twice the WHO-FAO "permissible rate." Again, the voluminous and carefully-documented background information indicates that no toxic effects have been detected or could be anticipated at this level, in babies, children, or adults. The levels of DDT encountered in various conditions and the effects of some of these levels are summarized in Table 1.

Antioxidants

Several workers have found that mice and rats fed diets with large amounts of added antioxidants live longer than those fed standard laboratory diets (17-20). The antioxidants include vitamin E, 2-mercaptoethylamine, ethoxyquin, butylated hydroxytoluene (BHT), and nordihydroguaiaretic acid (NDGA). The rationale for the effect is that natural and radiation-induced aging

TABLE 2   Chlortetracycline (CTC) in tissues of chickens*

CTC IN DIET	CTC IN TISSUE		
	Blood serum	Liver	Muscle
Experiment 1			
	p.p.m.		
0	0	0	0
200	0 -0.01	0	0
600	0.011-0.024	0 -0.10	0
2,000	0.039-0.024	0.14-0.30	0.05-0.16
Experiment 2			
0	0	0	0
2,000	0.06 -0.10	0.15-0.49	0.05-0.10
6,000	0.09 -0.35	0.18-0.39	0.08-0.33

*From Broquist and Kohler (22). Five replicate groups of 5 chickens each used at each level.

may be accompanied by a free-radical attack on lipids, including mitochondrial lipids, and this effect may be buffered by antioxidants (17). In a longevity experiment with mice fed a diet containing 0.5 per cent ethyoxyquin, the mice fed the antioxidant had an 18 per cent increase in average life span over the controls. (20). The authors suggest as one explanation that many antioxidants are powerful enzyme inducers; ethyoxyquin may cause liver enlargement. They suggest that "experiments with other potent enzyme inducers, such as DDT or barbiturates, seem to be necessary."

Antibiotics in Animal Foods

Antibiotics have been used successfully in feeding farm animals for twenty years. The procedure results in lower morbidity, lower mortality, more rapid gains in body weight, and an increase in the efficiency with which livestock feed is converted into human foods, such as meat and eggs. The value of antibiotics in feeds is due to the fact that domestic animals live in an environment that is universally contaminated with harmful microorganisms, many of which are susceptible to antibiotics.

The use of antibiotic supplements in commercial feeds was rapidly and widely adopted, starting in 1950 to 1952, as a means of improving the growth of poultry, pigs and calves (21). The levels used were low, usually 10 gm. or less of antibiotic per ton of feed. The practice gave satisfactory results over prolonged periods, and investigations soon turned to the use of higher levels in the treatment or prevention of certain endemic diseases of livestock.

Many members of the public are asking questions about antibiotic residues in their food. What quantities of such residues are present in food produced from domestic animals and what are their pharmacologic as well as their bacteriologic effects? These points will be discussed with chlortetracycline as an example, in view of its wide use in animal feeds.

The amounts present in chicken tissues following the feeding of chlortetracycline were measured by Broquist and Kohler (22) as summarized in Table 2. The antibiotic was not detectable in the muscle meat until a level was fed that is about ten times as high as any that is used continuously in practice. Additional experiments with an even more sensitive method using *Bacillus mycoides* spores showed that the muscle meat of chickens fed 50 p.p.m. chlortetracycline for twelve

121

TABLE 3 Chlortetracycline (CTC) residues in chicken tissues after oral administration for 5 days

TIME AFTER WITH-DRAWAL	AVERAGE CTC TISSUE CONCENTRATION			
	Muscle	Liver	Kidney	Fat
800 Gm. CTC per Ton in Diet				
days	←	— mcg./gm. —		→
0	0.38	0.90	6.44	0.09
1	0.01	0.02	0.16	neg†
3	neg†	0.01	0.12	neg†
2,000 Gm. CTC per Ton in Diet				
0	0.63	1.55	11.8	0.17
1	0.05	0.07	0.49	0.01
3	neg†-0.02	0.02	0.20	neg†
6	neg†	neg†	0.11	neg†

*From Shor *et al.* (23).
†No activity or less than 0.025 mcg. per gram.

weeks contained less than 1 part per 100 million of the antibiotic as a residue. More recent experiments by Shor, Abbey, and Gale (23) have produced essentially similar findings (Table 3). For example, chickens fed 2,000 gm. chlortetracycline per ton of feed, which is about ten times as much as is used in practice, had 0.63 p.p.m. antibiotic in their flesh. To get an average dose of chlortetracycline, a dose commonly prescribed for home use for patients by physicians, a person would have to eat 1 ton of raw chicken meat at a sitting! If the chickens were taken off feed for a day, the amount of raw chicken would have to be 10 tons, and if the meat were cooked, the antibiotic would be destroyed. Truly, the public in 1971 is being encouraged to get worried about nonsense.

The effect of cooking on chlortetracycline residues in beef, fish, and chicken meat was studied during extensive studies of chlortetracycline in the prevention of food spoilage. Broquist and Kohler found (22) that chlortetracycline disappeared from poultry meat on roasting (Table 4). The degradation product during cooking was identified as isochlortetracycline (24) which has an oral LD_{50}[2] in mice greater than 10 gm. per kilogram and has no known antibacterial effect.

For several years, chlortetracycline and oxytetracycline were used for delaying spoilage of poultry

[2]LD_{50} indicated level of dosage that is lethal to 50 per cent of the test group.

meat and fresh fish. The application to fish was especially useful in Japan and in Canada, where it was developed by Tarr at the Canadian Government Laboratories in British Columbia (25). No side effects of public health problems arose, but the FDA authori-

TABLE 4 Effect of roasting at 230°F. on chlortetracycline (CTC) in chicken breast muscle*

COOKING TIME	RESIDUAL CTC IN MUSCLE			
	0.03 mg. CTC/ml. dipping solution	0.1 mg. CTC/ml. dipping solution	0.3 mg. CTC/ml. dipping solution	1.0 mg. CTC/ml. dipping solution
min.	←————————p.p.m.————————→			
0	8.25†	25.0	60.0	120.0
15	none	none	0.33	1.15
30	none	none	none	0.31

*From Broquist and Kohler (22).

†A level of 8.25 p.p.m. CTC is about 13 times as high a level in meat as resulted from feeding CTC at 2,000 p.p.m. in the diet (23). This, in turn, is about 5 times as high as any level used in feeding animals.

zation for this use was withdrawn about three years ago, as a result of theoretical considerations raised by bacteriologists, who suggested that bacterial resistance might develop.

Tarr found (25) that heating salmon or lingcod flesh containing 5 to 10 p.p.m. chlortetracycline was followed by destruction of at least 80 to 90 per cent of antibiotic. Similarly Tomiyama et al. reported (26) that chlortetracycline disappeared from bonito fillets heated for 60 min. at 93°. The fillets originally contained 21 p.p.m. of the antibiotic.

Chlortetracycline is an unstable substance and decomposes fairly rapidly in meat even at refrigerator temperatures. Goldberg et al. found (27) that 2 p.p.m. added to ground beef disappeared after 96 hr. of storage at 10°.

Chlortetracycline (CTC) was in extensive commercial use as a food additive for a number of years in Canada for delaying spoilage in poultry meat. The raw meat was dipped in a solution of the antibiotic. This practice was valuable because it enabled the effects of residues to be studied at a higher level than that occurring in the meat of chickens fed the antibiotic as an animal feed ingredient. Thatcher and Loit of the Canadian Food and Drug Directorate carried out microbiologic examinations of commercial poultry meat samples, both untreated and treated with chlortetracycline, for various microbial categories (28). Salmonella was isolated from only one specimen of treated poultry meat, but was present in sixteen untreated samples. Four serotypes were recognized, including *S. typhimurium.* Cell isolates grew in broth containing 0.21 p.p.m. of CTC, but none grew at 0.43 p.p.m. Thatcher and Loit commented that "no data to indicate an increased health hazard as a result of the use of CTC as a poultry preservative were obtained. No evidence was revealed for modification of hazard due to the presence of staphylococci, enterococci, coliforms, or pathogenic yeasts."

The effects of prolonged feeding of antibiotics to human subjects were investigated during 1950 to 1955 (29). Reports in the medical literature reviewed up to 1956 described the long-term administration of chlortetracycline to 889 patients, all under close medical supervision, and drawn from all age groups. Approximately half were subjected to extensive clinical tests, including blood and bone marrow examinations and liver function tests. No evidence of toxicity from the antibiotic was found. In one study (30), two patients received 3 to 4 gm. chlortetracycline daily over twenty-one- and eleven-month periods, respectively. Initially, and every three months, liver biopsy and sternal marrow samples were examined. A peripheral blood study and urinalysis were done weekly, blood non-protein nitrogen monthly, electrocardiograms every two months, and also urinary ketosteroids, liver function tests, and glucose tolerance tests. No toxic manifestations were detected. It is unusual for a chemotherapeutic substance to be administered to human subjects for so long a period at so high a level without side effects.

Of particular interest are the results with infants and children. A total of 423 children received chlortetracycline, 20 to 500 mg. per day, by mouth for two to thirty-six months. In addition, there were findings on 120 premature infants who received 20 to 50 mg. per kilogram body weight per day for five to fifty-six days. Among the results in this group were lower mortality, more rapid gain in weight, and lower incidence of diarrhea. There were no reports of any signs of toxicity or of outbreaks of disease due to resistant pathogens (29).

In one study (31), chlortetracycline was given at 60 mg. per kilogram body weight per day. All treated infants survived, and five of the fifteen controls died.

In another investigation (32), in which 50 mg. chlortetracycline were fed daily, there was one death in forty-seven premature infants who received the antibiotic and eight deaths among the forty-eight controls.

Summary

The public reaction towards the use of chemical technology in the production and processing of food has been greatly heightened by the environmentalist movement. Much misinformation has been circulated in the news media. I have selected a few examples of residue problems for discussion, each illustrating a different point of importance.

The propaganda against DDT served to create distrust of all pesticides. Yet DDT is one of the safest compounds ever to be placed in contact with human beings. It has saved more lives and made more people healthy than any chemical in the history of the world. Its effects on wildlife are largely unknown, because wild animals pick up other contaminants from the environment, including polychlorinated biphenyls, lead, and mercury.

Of these, mercury is a natural ingredient of the ocean. According to Hammond (6), the total input of mercury into the seas from man's activities is between one-hundredth and one-thousandth of the hundred million tons of mercury in the ocean. He estimates that "except for coastal and estuarial areas, it does not seem likely that man could have increased concentrations in the sea by as much as 1 per cent." Clearly, the mercury present in fish of the deep oceans, including tuna, swordfish, sailfish, and albacore, is of natural origin. The Food and Drug Administration should make this clear to the public.

The addition of synthetic antioxidants to processed foods will no doubt draw the wrath of the natural food enthusiasts. Yet a typical antioxidant has been shown to prolong the life of mice to a highly significant extent.

The major antibiotics used in animal feeds were thoroughly tested for safety more than fifteen years ago. Their use has improved the health of farm animals and has increased the yields of animal products for human food, including beef, pork, poultry, and eggs. The residues in these products are a tiny fraction of the amounts of antibiotics used routinely in medicine, and the residues are destroyed by cooking.

These examples illustrate the tendency to exaggerate

125

and misinterpret the food residue problem. Adequate and accurate scientific information, especially on toxicology, is needed before decisions are taken.

References

(1) GUINEE, P.: Bacterial drug resistance in animals. Trans. N. Y. Acad. Sci., in press.
(2) CARSON, R.: Silent Spring. Boston: Houghton Mifflin Co., 1962.
(3) HAYES, W. J., JR.: Toxicity of pesticides in man. Proc. Roy. Soc. B 167: 101, 1967.
(4) The place of DDT in operations against malaria and other vector-borne diseases. WHO statement, EB 47/WP/14, 22 Jan. 1971.
(5) FRAZIER, B. E., CHESTERS, G., AND LEE, G. B.: "Apparent" organochlorine insecticide content of soil, sampled in 1910. Pesticides Monitoring J. 4: 67, 1970.
(6) HAMMOND, A. L.. Mercury in the environment: Natural and human factors. Science 171: 788, 1971.
(7) Tuna scare: Was pollution to blame? Marine Pollution Bull. 2: 2 (Jan.), 1971.
(8) How safe is your food? Interview with C. C. Edwards. U. S. News & World Rept. 70: 50 (Apr. 19), 1971.
(9) Report of the Secretary's Commission on Pesticides and Their Relation to Environmental Health. Pts. 1 & 2. Washington, D. C.: Dept. Health, Education, & Welfare, Dec. 1969.
(10) MILLER, J. A.: Tumorigenic and carcinogenic natural products. In Toxicants Occurring Naturally in Foods. Natl. Acad. Sci.-Natl. Research Council Pub. No. 1354, 1967.
(11) CHERKES, L. A., APTEKAR, S. G., AND VOLGAREV, M. N.: Hepatic tumors induced by selenium. Bull. Exp. Biol. Med. 53: 313, 1963.
(12) GARCIA-MARTIN, G. (WHO): Personal communication to S. Rotrosen, June 19, 1969.
(13) HAYES, W. J., JR.: Personal communication, 1970.
(14) RUSSELL, P. F.: The United States and malaria: Debits and credits. Bull. N. Y. Acad. Med. 44: 623, 1968.
(15) JUKES, T. H.: DDT: The chemical of social change. Clin. Toxicol. 2: 359, 1969.
(16) LAUG, E. P., KUNZE, F. M., AND PRICKETT, C. S.: Occurrence of DDT in human fat and milk. Arch. Industr. Hyg. 3: 245, 1951.
(17) PRYOR, W. A.: Free radical pathology. Chem. Engin. News, June 7, 1971, p. 34.
(18) HARMAN, D.: Free radical theory of aging: Effect of free radical reaction inhibitors on the mortality rate LAF mice. J. Gerontol. 23: 476, 1968.
(19) BUU-HOI, N. P., AND RATSIMAMANGA, A. R.: Action retardante de l'acide nordihydroguairèteque sur le vieillissement chez le rat. Compt. Rend. Soc. Biol. 153: 1180, 1959.
(20) COMFORT, A., YOUHOTSKY-GORE, I., AND PATHMANATHAN, K.: Effect of ethoxyquin on the longevity of C3H mice. Nature 229: 254, 1971.

(21) JUKES, T. H.: Antibiotics in Nutrition. N. Y.: Medical Encyclopedia, Inc., 1955.

(22) BROQUIST, H. P., AND KOHLER, A. R.: Studies of the antibiotic potency in the meat of animals fed chlortetracycline. *In* Antibiotics Annual, 1953-54. N. Y.: Medical Encyclopedia, Inc., 1954.

(23) SHOR, A. L., ABBEY, A., AND GALE, G. O.: Disappearance of chlortetracycline residues from edible tissues. 2. Chickens and turkeys. *In* Antimicrobial Agents and Chemotherapy. Washington, D. C.: Amer. Soc. for Microbiol., 1967, p. 757.

(24) SHIRK, R. J., WHITEHILL, A. R., AND HINES, L. J. A degradation product in cooked chlortetracycline-treated poultry. *In* Antibiotics Annual. N. Y.: Medical Encyclopedia, Inc., 1957, p. 843.

(25) TARR, H. L. A.: Control of Bacterial Spoilage of Fish with Antibiotics. Canad. Natl. Acad. Sci. Pub. 397, 1956, p. 199.

(26) TOMIYAMA, T., YONE, Y., AND MIKAJIRI, K.: Uptake of aureomycin chlortetracycline by fish and its heat inactivation. Food Tech. 11: 290, 1957.

(27) GOLDBERG, H. S., WEISER, H. H., AND DEATHERAGE, F. E.: Aureomycin in the prevention of spoilage of beef. Food Tech. 7: 165, 1953.

(28) THATCHER, F. S., AND LOIT, A.: Comparative microflora of chlortetracycline-treated and nontreated poultry with special reference to public health aspects. Appl. Microbiol. 9: 39, 1961.

(29) HINES, L. R.: Appraisal of effects of long-term chlortetracycline administration. Antibiot. Chemother. 6: 623, 1956.

(30) McVAY, L. V., AND CARROLL, D. S.: Aureomycin treatment of systemic North American blastomycosis. Amer. J. Med. 12: 289, 1952.

(31) ROBINSON, P.: Control trial of aureomycin in premature twins and triplets. Lancet 1: 52, 1952.

(32) SNELLING, C. E., AND JOHNSON, R.: Value of aureomycin in prevention of cross infection in hospital for sick children. Canad. Med. Ass. J. 66: 6 (Jan), 1952.

(33) PAL, R.: Contributions of insecticides to public health in India. World Rev. Pest Control 1: 6, 1962.

The Global "Cranberry Incident"

Thomas H. Jukes

In 1959, the announcement that certain batches of cranberries contained traces of aminotriazole, which was said to be a carcinogen, started a public panic which resulted in the destruction of hundreds of tons of cranberries. Aminotriazole is an anti-thyroid substance, and such substances produce thyroid tumors in rats when fed at high doses for prolonged times. The same effect results when rats are deprived of iodine in their food supply. There is an interesting parallel between the 1959 cranberry incident, and recent developments in the DDT question, which are having global repercussions rather than being confined to a cranberry scare.

The most serious indictment of DDT is the statement in the report of the HEW Commission on Pesticides [1] that if the Delaney clause were enforced, most food of animal origin would be outlawed because of traces of DDT. This says that DDT causes cancer. The statement is evidently based primarily on an article by Innes and co-workers [2] on tumors in the livers of mice that were fed a diet containing 140 ppm of DDT, described as "the maximum tolerated level." Similar tumors were repeatedly described in rats and mice during the period 1947 to 1959, and the scientific consensus was that the tumors, or nodules, were not cancerous because they disappeared when DDT was withdrawn from the diet. They were not produced in monkeys, dogs, cats, chickens, or large domestic animals dosed with DDT [3]. Hayes [4] comments on the Innes article [2] as follows:

"The incidence of tumors was comparable to the mean tumor incidence produced by a group of positive control compounds, most of which are weak or even questionable carcinogens of no demonstrated importance to human health. The authors made no distinction between hepatomas and carcinomas. It is difficult to understand why, in denying the practicality of making this important distinction, they entirely neglected the matter of reversibility . . . In the meantime there is no assurance that the small number of tumors observed in mice exposed to DDT were different from the 'nodules' described by Fitzhugh and Nelson [5]. Furthermore, the entire testing scheme was adopted in the hope of achieving greater sensitivity, but no responsibility has been taken for measuring its biological significance. There is no assurance that the same test would not give positive results for some common items of the diet such as spices, caffeine, or even table salt . . . Unless more convincing evidence is obtained than that reviewed above, I conclude that Dr. Lehman [6] was correct. DDT is not a carcinogen."

The effect of the statement by the HEW Commission has been to set in motion a wave of public fear that DDT residues in the food will cause cancer. The director of the crop protection branch of the Food and Agriculture Organization (FAO) stated that "with press reports tending to give the impression that DDT causes cancer, there is going to be no minister of health who can stand up and defend its continued use."

The rumor of carcinogenicity has been skillfully abetted by those who seek a ban on DDT. A "global cranberry incident" is in full swing; unfortunately it will have tragic consequences for the future victims of malaria.

All ideas of quantities have been cast to the winds. The results with mice [4] were obtained at a level which corresponded to about 3000 times the average U.S. human exposure on a food intake basis or about 40,000 times on a body weight basis. The effect of DDT was to produce an eightfold increase over the controls in the incidence of liver nodules. On a linear dose-response basis, and if the nodules were cancerous (which they were probably not), this would correspond to about one additional case of liver cancer in a population of 200 million human beings having 4000 cases annually, if the calculation is based on body weight, or eight cases on a food intake basis. However, it appears likely that DDT may prevent liver cancer produced by a "natural" carcinogen (see below).

It is paradoxical and ironical that well-authenticated work shows DDT can have an anticarcinogenic effect by inducing the formation of liver enzymes that destroy aflatoxin [27]. Aflatoxin may be a cause of endemic cancer in human populations whose principal supply of protein comes from peanuts, as in East Africa. Rats were protected against death from aflatoxin by dosing with DDT [8]. Other work on DDT and experimental cancer is in progress which indicates that DDT increased by 60% the life span of mice that were implanted with methylcholanthrene-induced ependymoblastomas [9].

And now, let us sit back and watch the genocidal consequences of the "global cranberry incident" set in motion by the press conference held in November, 1969 by the Secretary of Health, Education, and Welfare.

REFERENCES

[1] *Secretary's Commission on Pesticides and Their Relation to Environmental Health. Part I. Recommendation and Summaries,* Department of Health, Education and Welfare, November 1969, pp. i-xvii and 1-44.

[2] J. R. M. Innes, B. M. Ulland, M. G. Valerio, L. Petrucelli, L. Fishbein, E. R. Hart, A. J. Pallotta, R. R. Bates, H. L. Falk, J. J. Gart, M. Klein, I. Mitchell and J. Peters, *J. Natl. Cancer Inst.,* 24, 1101 (1969).

[3] W. J. Hayes, Jr., in *DDT. The Insecticide Dichlorodiphenyltrichloroethane and its Significance, Vol. II, Human and Veterinary Medicine* (P. Müller, ed.), Birkhauser Verlag, Basel, 1959, pp. 11-247.

[4] W. J. Hayes, Statement made at hearings on DDT, Seattle, Washington, October 15, 1969.

[5] O. G. Fitzhugh and A. A. Nelson, The Chronic Oral Toxicity of DDT (2,2-bis (p-chlorophenyl)-1,1,1-trichloroethane), *J. Pharmacol.,* 89, 18-30 (1947).

[6] A. G. Lehman, *Summaries of Pesticide Toxicity,* The Association of Food and Drug Officials of the United States, 1965.

[7] A. E. M. McLean and E. K. McLean, *Brit. Med. Bull.,* 25, 278 (1969).

[8] A. E. M. McLean and E. K. McLean, *Proc. Nutr. Soc.,* 26, xiii (1967).

[9] E. R. Laws, personal communication (1969).

DDT In The Ecosystem

DDT in the Biosphere:
Where Does It Go?

George M. Woodwell, Paul P. Craig, and Horton A. Johnson

DDT has been used in large quantities as an insecticide since 1942. Its residues (*1*) are sufficiently persistent and mobile to have a worldwide distribution, appearing in the lipids of most organisms (*2–7*), in air (*7–9*), and occasionally in meltwaters of Antarctic snows (*10*). Concentrations in certain of the earth's biota have reached toxic levels, causing spectacular declines in populations of certain carnivorous and scavenging birds and fish, aggravating the problems of pollution, and threatening significant contamination of human food chains (*2, 4, 11–14*). Recognition of the seriousness of these problems has led to recent restrictions in the use of DDT in the United States and abroad. There is at least a possibility that most of the DDT that has been or will ever be produced has already been used and that little, if any, will be applied after the mid-1970's (*15*). The persistence of DDT residues is great enough, however, that the residues will continue to be redistributed for many years after use of the pesticide has stopped, presumably presenting a continuing hazard to all the biota. The extensive data available on the distribution and effects of DDT make it, together with radioactive substances, the best known of the biospheric pollutants and a valuable subject for a case history study (*12, 16, 17*). But basic questions remain, among them the following: What becomes of DDT released into the biosphere? How serious are the hazards? and, How long will the hazards persist?

In an effort to answer these questions we have attempted to develop a model of the circulation of DDT in the biosphere. We have done this on the basis of two limiting assumptions: (i) that use of DDT will decline to zero by 1974 and, alternatively, (ii) that, between now and then, use will increase throughout the world.

Certain physical properties of DDT are important in determining its behavior in the biosphere. First, because of the high solubility of DDT in fats together with its low solubility in water (*18, 19*), DDT residues tend to accumulate in lipids and therefore in plants and animals. Second, the residues are very persistent in nature: estimates of their half-life range upward to 20 years, perhaps longer under certain circumstances (*20–24*). Third, DDT has a vapor pressure high enough to assure direct losses from plants and soil into the atmosphere, which can carry residues worldwide (*12, 13*). Thus soils, air, the waters of the oceans, and the biota are all potential reservoirs for DDT residues, and the hazard to the biota, including man, hinges on the distribution of DDT residues among these reservoirs. How large are the reservoirs, and what are the rates of exchange between them?

The answers are not available in any simple or absolute sense. Most are available, however, at least by inference. First, we must know how much DDT has been produced and something about its distribution.

Input: DDT Production

The amount of DDT produced in the United States each year is reported by the U.S. Tariff Commission (*25*). In the crop year 1963 the amount of DDT produced reached a maximum 8.13 × 10^{10} grams (179 × 10^6 pounds) (Fig. 1). Production has dropped in the United States since 1963, but more than 6.0 × 10^{10} grams of DDT were produced in 1969. Preliminary figures for 1970 reveal that DDT production declined by more than 50 percent. About 70 percent of the amount of DDT produced appears to have been used outside the United States. The total amount of DDT produced in the United States, integrated over the entire period through 1974, when we have assumed that DDT will no longer be used, is estimated to be 1.4 × 10^{12} grams. No data on world production are available. We have assumed that the amount of DDT produced in the world, including the U.S. fractions, is twice the amount produced in the United States, or about 2.8 × 10^{12} grams in total through 1974.

This DDT has been distributed widely around the world, most heavily in humid temperate and tropical zones. It is commonly applied by spraying a liquid suspension or solution from mobile ground equipment or from aircraft. The fraction of the spray that lands on the target varies, but some fraction of both aerial and ground applications remains airborne. Aerial applications of DDT to forests in the northeastern United States show that 50 percent or less of the amount emitted from the planes is deposited on the forest. The rest is dispersed into the air. Much of the airborne fraction returns to the ground nearby, but small droplets or particles are likely to remain aloft, to become associated with other particles, and may be carried great distances (*7–9, 14, 26*).

DDT in Soils

The persistence of DDT led to early recognition that residues might accumulate in soils. On the basis of a review of published data (Table 1), we have estimated that agricultural soils in the United States contain an average content of DDT approaching 0.168 gram per square meter (1.50 pounds per acre) (*22, 23, 27–35*). Nonagricultural soils were estimated to contain an average of 4.5 × 10^{-4} gram per square meter (*36*).

These estimates can be used to calcu-

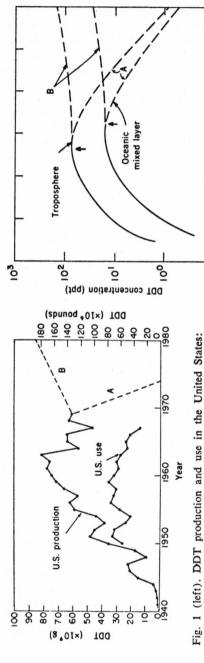

Fig. 1 (left). DDT production and use in the United States; curve A, based on the assumption of declining use through 1974; curve B, based on the assumption of increasing use through 1980. Dotted lines indicate projections. Fig. 2 (right). DDT concentrations in the mixed layer of the oceans, in the troposphere, and in the biota projected through the year 2000 on the basis of assumptions A and B of Fig. 1.

134

late the rates of loss of DDT residues from soils in the United States. Within the United States, the total contiguous land area is 7.7×10^{12} square meters (1.9×10^9 acres), about 11 percent of which (0.85×10^{12} square meters) is kept in crops on which insecticides are used (37). The agricultural land apparently retains about 1.42×10^{11} grams of DDT. These data were obtained in the early 1960's when DDT production was high. The rate of use of DDT in the United States during this period was 2.7×10^{10} grams per year (6×10^7 pounds per year) (Fig. 1) or 0.0318 gram per year for each square meter of agricultural cropland (0.28 pound per acre per year). The mean lifetime of DDT in soils must be about $10^{11}/(2.7 \times 10^{10}) = 5.3$ years. This estimate approximates Edwards' earlier estimate of 4.3 years (21, 38); it is substantially less than the estimated lifetime of 10 years for DDT in certain soils (22, 24). We have assumed a mean resident time of 4.5 years for DDT on land.

Four mechanisms probably account for most losses of DDT residues from soils: (i) volatilization (including losses by wind erosion of small particles from the soil surface), (ii) removal by harvest of organic matter, (iii) water runoff, and (iv) chemical (including biotic) degradation.

The occurrence of DDT residues in rainwater suggests that large amounts of DDT may move through the atmosphere either adsorbed to particles or as vapor. The small number of data available on the volatilization of DDT suggest a time constant for volatilization of several years, but the evidence supports the conclusion that vaporization is more important than such a long time constant would indicate (14, 21, 38, 39). The rates of disappearance of residues of dieldrin (1) by volatilization have been shown to be a function of the rate of air movement through soils (40). Residence

Table 1. DDT residues in soils of agricultural and nonagricultural land in the United States. Data were selected because of large sample size or because they are the only data available. For more detailed tabulations see Edwards (21, 38).

Soil sites	Sites sampled (No.)	DDT residues (g/m²)		Reference
		Range	Mean	
Agricultural				
Orchards	14	0.34 –22.1	6.0	(29)
Crops	24	0 – 0.87	0.24	(29)
Root crops	48	0.045 – 5.73	1.25	(30)
Vineyards	2	2.13 – 3.18	2.69	(31)
Orchards	2	8.18 –14.60	11.4	(32)
Vegetable crops	10	0.07 – 9.52	2.62	(33)
Randomly selected	41	0.002 – 1.30	0.148	(33)
Alfalfa crops	12	0.06 – 0.98	0.336	(34)
Soybean crops	43	0.004 – 4.03	0.986	(35)
Nonagricultural				
Boreal forests (sprayed)	3	0.179 – 0.258	0.213	(22)
Boreal forests (unsprayed)			0.0045	(23)
Forest in Pennsylvania (unsprayed)		0.0003– 0.0006	0.0004	(28)

times for DDT in organic soils where movement is slow are greater than residence times in mineral soils, a relationship that could only be the case if biotic degradation of DDT residues proceeds slowly in these soils in comparison with volatilization. Apparently evaporation is a major mechanism for the removal of residues of the persistent pesticides from soils (21, 38), and, despite its slowness, evaporation proceeds faster than chemical breakdown.

DDT residues are also removed from soils by the harvest of organic matter, but the evidence suggests that this is not a major route of transport. The net amount of crop harvested, even for highly productive agricultural crops, seldom exceeds 5000 grams per square meter and more often approaches 1000 grams per square meter (41). If we assume an annual harvest of 3000 grams per square meter containing 1 part per million (ppm) of DDT, currently the maximum concentration allowed in many foods, the harvest would remove 0.003 gram per square meter of DDT, about 1 percent of an annual application of 0.33 gram per square meter, or about 10 percent of the estimated annual average amount of DDT used for all crop land (see above). Thus, even a harvest of 100 percent of the primary production would remove only a small fraction of an annual application of DDT. Actual harvests that remove the organic matter from the site and might transport DDT residues to urban areas or to watercourses are much less efficient. We have assumed a removal of 1 percent of the total DDT used on the crop.

DDT is not normally applied to forage crops or in places where it can contaminate tissues of animals used for food, but farm animals do, nonetheless, become contaminated, as do most animals. The amount of farm animal biomass produced each year usually approximates 10 percent or less of the net primary production, and it is difficult to visualize under present levels and patterns of use a circumstance in which harvest of farm or range animals would account for more of the DDT residues than would the annual harvest of the plant crops, despite the possibility of concentration of the residues through food chain effects.

The transport of DDT residues from agricultural soils in surface waters has often been assumed to be of major importance in the accumulation of residues in the oceans. Heavy rains do remove DDT either adsorbed to soil particles or in solution but the amounts are small in comparison with the total amounts of DDT produced. Surface runoff over the entire United States is about one-third of the annual precipitation, or 23 centimeters (9 inches) per year out of an average precipitation estimated as 76 centimeters (42). The total volume of this water is 2×10^{12} cubic meters per year. If it were saturated with DDT residues, it would contain about 1 part per billion (ppb) (18, 43) and would remove 2×10^9 grams of DDT per year in solution. DDT is applied only at certain times of the year in certain areas and seldom directly to bodies of water. Observed river water concentrations of pesticides (including particles) range from concentrations below the limit of detection [less than 10 parts per trillion (ppt)] to almost 100 ppt (18, 43). An average concentration of 50 ppt implies an annual runoff of about 10^8 grams per year, which accounts for about 0.1 percent of the amount of DDT produced per year. Similar conclusions have been reached by Risebrough et al. (9), who also decided that movement of residues in the atmosphere is the most important transport route, and by the ocean pollution group of the M.I.T. Study of Critical Environmental Problems (14).

DDT in the Atmosphere

The vapor pressure of DDT at 20°C is 1.5×10^{-7} millimeter of mercury (44), producing an equilibrium concentration of DDT in the atmosphere of about 3×10^{-6} gram per cubic meter or about 2 ppb by weight. The vapor pressure drops with decreasing temperature. If we assume that DDT in the atmosphere remains as vapor, the saturation capacity of the atmosphere to the tropopause would be about 10^{12} grams of DDT, or about as much as has been produced to date. But DDT residues also exist in association with atmospheric particles, and the earth's atmosphere can probably contain very much larger quantities than the saturation capacity alone would indicate. This means that the atmosphere is potentially a large reservoir in addition to being a major means of transport for the residues.

Residues are removed from the atmosphere by rainfall, diffusion across the air-sea interface, and chemical degradation. The dominant mechanism for the removal of DDT from the atmosphere is probably rainfall. In England, DDT concentrations in the range from 73 to 210 ppm have been reported in rain in areas close to regions where DDT has been used, and similar concentrations have been reported in the United States (45). A DDT concentration in meltwaters from Antarctic ice of 40 ppt has been reported recently (10). Earlier measurements were less sensitive (46).

DDT concentrations in rainfall vary appreciably throughout the year. The variation is related to the seasonal application of DDT. The fact that there is a seasonal variation suggests that the time constant for removal of DDT from the atmosphere probably does not exceed a few years. If the average DDT concentration in rainfall were 60 ppt and precipitation averaged 1 meter per year, rainfall would remove a total of 3×10^{10} grams of DDT residues from the atmosphere annually, most of that into the oceans. The annual amount of DDT produced throughout the world in the mid-1960's was about 10^{11} grams, approximately 3½ times the amount that would be removed worldwide by rain containing 60 ppt of DDT. Thus an average rainfall concentration of DDT of 60 ppt gives an upper limit for the mean time for removal of DDT residues from the atmosphere by rainfall of about 3.3 years. Residues deposited on the ground are, of course, available for reevaporation.

Measurements of the transfer of carbon dioxide into the oceans suggest a time constant for the downward transport of DDT of 7 years (47, 48). This period is extremely long as compared to estimates for diffusive transport. Although one might expect atmospheric DDT to be transferred more slowly than carbon dioxide because of DDT's low solubility in water, direct measurements seem to be lacking. Accumulation of lipids at the ocean surface (49) would probably increase the rate of transfer as would association of DDT residues with particles in air. We have assumed a time constant of 4 years for the transfer of DDT residues from the atmosphere to the earth's surface. The mean residence time is probably not longer than this.

Chemical degradation of DDT in the atmosphere may also be important. Efficient photodegradation of DDT vapor occurs primarily at wavelengths shorter than 2700 angstroms (50). These wavelengths are heavily absorbed by the atmospheric ozone layer. Residues adsorbed on particles are probably considerably more resistant, however. We have assumed atmospheric degradation to be unimportant as compared to transport, but this topic is obviously in need of further study.

DDT in the Oceans

DDT residues circulate initially in the "mixed" layer, which frequently extends to a depth of 75 to 100 meters. They are transferred slowly below the thermocline into the much larger volume of the abyss (51). Sedimentation of organic matter removes DDT residues from the upper layers, but direct measurements of sedimentation of DDT residues seem to be lacking. We assume that the virtual insolubility of DDT in water combined with its solubility in fat assures the association of DDT with organic matter and that the rate of transfer of carbon to the abyss would approximate the rate of transfer of DDT residues. Biological mixing may also be important in DDT transport within the ocean, but direct evidence is lacking. As an estimate of the time for the transport of DDT from the mixed layer to the abyss, we have used a result from studies of carbon dioxide that indicate a mean transfer time of about 4 years (47, 48).

Within the abyss, transfer rates for carbon dioxide and presumably for other substances such as DDT are very slow indeed, ranging up to hundreds and even thousands of years for certain segments. Because of the fact that the volume of the abyss is immense and because of the possibility that DDT residues may be lost to sedimentation, the abyss is a very large reservoir, virtually infinite for the purposes of this discussion.

DDT in the Biota

The total amount of DDT retained within the biota is small by comparison with the totals that can be retained in other pools within the biosphere. It is also small by comparison with the annual amount of DDT produced. Liberal assumptions with respect to the concentrations of residues in various segments of the biota, including man, lead to an estimate of 5.4×10^9 grams of DDT held within the biota worldwide (Table 2). Estimates of the world biomass are notoriously variable but they are probably correct to within a factor of 2 to 3, almost certainly to within less than a factor of 10. DDT analyses that are specifically appropriate for compilation of such an inventory are few, and the data of Table 2 are, at best, crude estimates. The data on residues have been expressed to orders of magnitude only to avoid a false indication of precision. Questionable estimates have been resolved in the direction of the higher order of magnitude, thus giving a bias toward a higher estimate. The analysis indicates that there may now be between 10^9 and 10^{10} grams of DDT circulating in the biota, about 1/30 of the amount produced in 1 year during the mid-1960's. We consider this an estimate of the maximum amount of DDT that could be in the biota; the only other estimate available is 6×10^8 grams for the biota of the oceans alone published by the M.I.T. study group (14). This means that, despite the importance of the biota and the effects

138

of DDT on it, the capacity of the biota for holding DDT residues is small enough that we can ignore it for the moment in our attempt to appraise the worldwide movements of DDT.

A Model of DDT Circulation in the Biosphere

The number of pathways that are clearly important in the worldwide movement of residues appears to be small. The primary reservoirs are the land surface, the troposphere, the mixed layer of the ocean, and the abyss. In the previous sections, we have estimated time constants for the dominant physical processes. The constants have been used in a set of first-order rate equations to yield estimates of DDT loads in the various reservoirs as functions of time. The rate equations have the form:

$$\frac{dN_i}{dt} = R_i(t) - \sum_{j=1}^{m} \frac{N_i}{\tau_{ij}^{(l)}} +$$

$$\sum_{j=1}^{m} \frac{N_i}{\tau_{ij}^{(g)}}, i = 1 \ldots m$$

Table 2. DDT residues in the biota in the late 1960's. Concentrations are expressed to the nearest order of magnitude only; ppm, parts per million.

Location	Dry biomass ($\times$ 10⁹ metric tons)	DDT content (ppm)*	Total DDT ($\times$ 10⁸ g)
Plant biomass†			
On land			
Lakes and streams	0.04	0.010	0.004
Swamps and marshes	24	0.001	0.240
Terrestrial vegetation (forests, desert, savanna, grassland, tundra)	1814	0.0001	1.814
Agriculture	14	0.1	14.000
Total land	1852		15.058
In oceans			
Open ocean algae	1.0	0.1	1.0
Continental shelf algae	0.3	1.0	3.0
Attached algae	2.0	1.0	20.0
Total ocean	3.3		24.0
Total plants	1855		39.06
Animal biomass‡			
On land			
Feral mammals	0.009	1.0	0.09
Domestic mammals	0.17	1.0	1.7
Man	0.30§	1.0	3.0
Birds	0.00024	1.0	0.002
In oceans			
Fish	0.65	1.0	6.5
Mammals	0.055	1.0	0.55
Others (protozoa, coelenterates, annelids, nematodes, mollusks, echinoderms, arthropods)‖	3.02	0.1	3.02
Total animals	4.20		14.86
Total DDT in the biota:		5.4 $\times$ 10⁹ g	

* Estimates based on values in the literature and the experience of G.M.W. Sources include (2–7, 11–14, 20–24, 36, 55, 56) and others. All estimates are to the nearest order of magnitude only; questionable data have been resolved toward the higher number.　† Adapted from Whittaker (60). ‡ Adapted from Bowen (61).　§ Bowen used 0.03 × 10⁹ tons, which seems to be about 10 times too low (G.M.W.).　‖ Listed by Bowen (61); obviously incomplete, but indicative.

Here $R_i(t)$ is the rate at which newly produced DDT is introduced into the ith reservoir and N_i is the amount of DDT in the ith reservoir. The sums represent losses from and gains to the ith reservoir; $\tau_{ij}^{(l)}$ is the time constant for DDT loss from the ith to the jth reservoir, and $\tau_{ij}^{(g)}$ represents inputs from the jth to the ith reservoir. There are m reservoirs, and there are thus m simultaneous, first-order, differential equations to be solved.

We have used the time constants estimated above to solve the rate equations over the period from 1940 to 2000 on a digital computer. The DDT input for each year has been taken to be twice the amount produced in the United States per year and has been projected in two ways (Fig. 1, curves A and B). Except for the distinction between land and ocean, no attempt has been made to include geographical variation. Local fluctuations may be expected to be large.

The calculated average DDT concentrations in the atmosphere and in the mixed layer of the oceans for the period from 1940 to 2000 are shown in Fig. 2. If the world DDT production becomes zero in 1974, the concentration in the lower atmosphere would have reached a peak in 1966 at about 72 ppt (84×10^{-9} gram per cubic meter). The mixed layer of the ocean would contain its maximum of 15 ppt in 1971. The concentrations in both reservoirs can be expected to decline gradually, with the concentration in air reaching 10 percent of its peak value in 1984. The concentration in the mixed layer will not decline to 10 percent of its peak value until 1993. The total load of DDT on the land surface reached a maximum concentration of 6.34×10^{11} grams in 1964 to 1966, and will decline if DDT production slows.

There is reason to assume that the worldwide production of DDT will not drop, but may increase, despite U.S. restrictions. The increase will be in response to an increasing demand for an inexpensive means of pest control in agriculture and for control of vectors of disease, especially malaria. If we assume that foreign production replaces U.S. production and that the total use of DDT in the world increases after 1969 (curve B, Fig. 1), the concentrations of DDT in air and water will follow the curves marked B in Fig. 2, continuing to increase until after the year 2000.

Implications for Life

The physical processes we have discussed dominate the transfer of DDT residues throughout the biosphere. Living systems retain quantities of DDT that are small by proportion, and living things appear, at least superficially, to play a minor role in the world budget of DDT. Yet it is the residues that are available to living systems that are the hazard, and we must examine their behavior with special care. The total quantities of DDT residues in the biota are but 1/30 or less of the annual amount of DDT produced in recent years; they are also a small fraction of the annual transfers estimated from soils to air and the oceans. The quantities are small enough that the transfer from land to water by surface runoff, small as it is, must be assumed to contribute to contamination of the coastal biota. The residues presently held in the biota, and the maximum quantity that the biota could hold (not very greatly different), are so small in proportion to the total amount of DDT produced that we wonder why the biota has not been affected much more drastically than it has been—and what the future holds.

The answers are far from clear. DDT residues are accumulated in living systems and recycled in much the same ways that certain elements essential for life are recycled. Just as phosphorus is recycled from sediments by various means, so DDT residues may reenter complex food webs from organic sediments. One such route is by direct consumption of detritus (52). Others must include oxidation of the sediments. Concentrations of DDT in the biota may ultimately reach as much as 10^6 times the concentrations in the general environment (36). The effects of high concentrations are clear enough: food webs are reduced, carnivores eliminated, and hardy, small-bodied organisms favored (53). The changes are similar to those that occur in eutrophication; the sedimentation of organic matter is probably increased, often speeded by a shift toward an increasingly anaerobic benthos. We assume that under such extreme conditions DDT residues tend to accumulate in anaerobic sediments and are removed from circulation. Thus one of the effects of DDT is to reduce the biota and to increase the rate of removal of DDT residues into sediments. The process tends to restrict the movement of residues in the larger circulations of the biosphere, accentuating the importance of local contamination. There is, however, not much question that the oceans, as well as lakes and estuaries, are vulnerable to such effects. How much more DDT would it take to degrade the biota significantly?

The answer hinges on both the rate of movement of residues through the major reservoirs of the biosphere and on the coupling between the biota and the environment. How rapidly does the biota absorb DDT? Is there a possibility of the biota's achieving an equilibrium in which inputs of DDT residues are exactly balanced by losses?

DDT residues enter the biota both through food webs and by direct absorption. The relative importance of these routes varies between land and water and among species, and the time for the biota to come to equilibrium with residues in the environment must also vary. One attempt at appraising the time for a food web to reach equilibrium led to an estimate of between "four times the average life span of the longest-lived species and the sum of the life spans for all trophic levels" (54, p. 506). Such an analysis suggests that equilibrium for the entire biota would be reached only after many decades. Movement of DDT residues into the abyss appears much more rapid than this.

On the other hand, plankton in water would be expected to reach equilibrium with residues in solution in the water very rapidly, and small-bodied, warm-blooded carnivores that have high rates of metabolism and feed from water-based food webs might be expected to accumulate high concentrations of residues rapidly and to be affected by them. This circumstance, of course, is what we see: aquatic carnivorous birds accumulate high concentrations of DDT and then their numbers rapidly diminish. So, although the biota as a whole may not have achieved equilibrium with the residues circulating now, certain segments of the biota are being reduced, a fact that indicates that the biota may be appreciably changed before an "equilibrium" with present rates of DDT production and present world inventories is reached. Under these circumstances the concept of an equilibrium becomes elusive; there is no true equilibrium, only a constant state of flux through a pool that probably grows smaller as the biota is reduced. The coupling between DDT residues in the environment and in aquatic food webs would seem to be reasonably close, especially for lower trophic levels, and a decline in the amount of DDT used should be reflected almost immediately in lower concentrations in the biota.

A variety of evidence favors this conclusion. Fauna of salmon streams in the Miramichi River in New Brunswick, Canada, and fish in Sebago Lake, Maine, responded year by year to a reduction in the use of DDT (55). More recent experience on Long Island with osprey populations seems to be indicating a similar response, although the observations are far from conclusive. Populations dropped abruptly during the mid-1960's. The decline seems to have been arrested, perhaps reversed, after cessation of use of DDT for mosquito control in 1967 (56). Terrestrial ecosystems probably respond more slowly, but the patterns of circulation of residues and their effects are similar.

These examples and others, such as the observation that the effects of pesticides on reproduction of bird populations in England are related to the intensity of use of chlorinated pesticides (57), emphasize that most conspicuous effects on the biota in the past have not been the result of worldwide movements of DDT residues but rather have been attributable to local concentrations identifiable with some local or regional use. Restriction in the use of pesticides has usually reduced the effects within a few years.

Residues in other segments of the biota, however, are very much more closely related to larger patterns of circulation. Oceanic birds such as the sooty and slender-billed shearwaters (4, 5) and the Bermuda petrel (6) that feed in the open ocean, the California mackerel, the penguin, and the crabeater seal of the Antarctic (58) must obtain their residues from patterns of circulation that are close to being "worldwide." All of these organisms are contaminated with DDT, some with concentrations occasionally exceeding 10 ppm. These are the organisms that are most closely coupled to the major nonliving pools of DDT circulating in the biosphere. Our analyses suggest that such organisms should reflect world use of DDT within a few years, with residues in the biota increasing or decreasing as world use rises or falls. The fact that none of these organisms has yet become extinct from DDT effects is mere good fortune: the total amounts of DDT estimated to be circulating in the biosphere are many times greater than the amounts required to eliminate most such animals. We know that the residues can be concentrated by many factors of 10 into the biota—and affect it catastrophically. Yet, although there is no question about the devastation wrought by DDT locally and even regionally, the worldwide component seems not yet to have reached the point of widespread extinctions. (The difficulties of measuring effects are so great as to make most biologists who examine this question in any depth suspicious that effects may be occurring unobserved or masked by other causes). If we assume that the lower trophic levels of aquatic food webs are more closely coupled to their environment by dint of the two pathways for entry of residues than the analyses of Harrison et al. (54) suggest, then reduction in the use of DDT should be reflected within a few years in a reduction in the DDT residues in the biota identified with the worldwide distribution.

Where Has the DDT Gone?

The physical and chemical characteristics of DDT might lead one to assume that the biosphere should behave as a giant separatory funnel, gradually partitioning the lipid-soluble residues into the lipid-rich biota. Although there is no question that this process does occur, there is also no escape from the conclusion that it does not work well on the biospheric level. Most of the DDT produced has either been degraded to innocuousness or sequestered in places where it is not freely available to the biota. Recent work seems to support the latter assumption and the assumptions of our model. A preliminary report of detailed analyses of DDT residues in the air of nine U.S. cities in 1967–1968 (59) shows concentrations in winter, when DDT is not used locally and residues might be expected to be mixed throughout the troposphere, commonly falling between 10^{-9} and 100×10^{-9} gram per cubic meter. The range approximates our prediction of 84×10^{-9} gram per cubic meter in air based on the assumption of declining use. The observation supports our assumptions on the routes of movement and sizes of pools. The fact remains, however, that, despite the abundance, persistence, and worldwide distribution of DDT residues, they are not as freely available to the biota as might be assumed. How and precisely where they are held is not yet clear, but the biosphere appears to have a large capacity for holding them apart from the biota. What is clear is that large quantities of DDT were introduced into use before any appraisal was made of the capacity of the biosphere for receiving them. In this instance man seems to have been blessed with extraordinary good fortune.

Summary

The worldwide pattern of movement of DDT residues appears to be from the land through the atmosphere into the oceans and into the oceanic abyss. Calculations based on the fragmentary data available on rates of movement and sizes of various pools of DDT residues lead to the conclusion that concentrations in the atmosphere and in the mixed layer of the oceans lag by only a few years behind the amounts of DDT used annually throughout the world. A model suggests that maximum concentrations of DDT residues occurred in air in 1966 and will occur in the mixed layer of the oceans in 1971. The biota probably contains in total less than 1/30 of 1 year's production of DDT during the mid-1960's, a very small amount in proportion to the total potentially available. The reason for the biota's failure to absorb larger quantities and to be affected much more severely is unclear. The analysis suggests that mere good fortune has protected man and the rest of the biota from much higher concentrations, thus emphasizing the need to determine the details of the movement of DDT residues and other toxins through the biosphere and to move swiftly to bring world use of such toxins under rational control based on firm knowledge of local and worldwide cycles and hazards.

References and Notes

1. DDT residues include DDT and its decay products, DDD and DDE: DDT, 1,1,1,-tri-chloro-2,2-bis(p-chlorophenyl)ethane; DDE, 1,1 dichloro-2,2-bis(p-chlorophenyl)ethylene; DDD, 1,1-dichloro-2,2-bis(p-chlorophenyl)ethane; di-

eldrin, 1,2,3,4,10,10-hexachloro-6,7-epoxy-1,4,-4a,5,6,7,8,8a-octahydro-endo-exo-1,4:5,8-dimethanonaphthalene.

2. For a discussion of residues in man, see U.S. Department of Health, Education and Welfare, *Report of the Secretary's Commission on Pesticides and Their Relation to Environmental Health* (Government Printing Office, Washington, D.C., 1969).

3. Some of the most recent studies establishing the extent of contamination are: R. W. Risebrough (4); ———, D. B. Menzel, D. J. Martin, H. S. Olcott (5); C. F. Wurster, Jr., and D. B. Wingate (6); E. C. Tabor (7).

4. R. W. Risebrough, in *Chemical Fallout*, M. W. Miller and G. G. Berg, Eds. (Thomas, Springfield, Ill., 1969), p. 5.

5. ———, D. B. Menzel, D. J. Martin, H. S. Olcott, *Nature* 216, 589 (1967).

6. C. F. Wurster, Jr., and D. B. Wingate, *Science* 159, 979 (1968).

7. E. C. Tabor, *Trans. N.Y. Acad. Sci.* 28, 569 (1966).

8. P. Antommaria, M. Corn, L. DeMaio, *Science* 150, 1476 (1965); D. C. Abbott, R. B. Harrison, J. O. Tatton, J. Thompson, *Nature* 211, 259 (1966).

9. R. W. Risebrough, R. J. Huggett, J. J. Griffin, E. D. Goldberg, *Science* 159, 1233 (1968).

10. T. J. Peterle, *Nature* 224, 620 (1969).

11. N. C. W. Moore, Ed., *J. Appl. Ecol.* 3 (Suppl.), xxx (1966).

12. G. M. Woodwell, *Sci. Amer.* 216, 24 (March 1967).

13. J. Frost, *Environment* 11, 14 (1969).

14. Massachusetts Institute of Technology, *Man's Impact on the Global Environment: Report of the Study of Critical Environmental Problems* (M.I.T. Press, Cambridge, Mass., 1970), pp. 126–131.

15. In the United States the Secretary of Health, Education, and Welfare, the Secretary of Interior, the Secretary of Agriculture, and the administrator of the Environmental Protection Agency have moved recently toward removing DDT from most uses by 1972. These moves are being challenged, and the question remains unresolved. Certain other countries, including Canada and England, have restricted use of persistent pesticides. There is no basis for expecting an international ban and little basis for expecting restraint in the use of DDT outside of these countries.

16. G. M. Woodwell, *BioScience* 19, 884 (1969).

17. B. Commoner, *Science and Survival* (Viking, New York, 1963).

18. M. C. Bowman, F. Acree, Jr., M. K. Corbett, *Agr. Food Chem.* 8, 406 (1960).

19. T. F. West and G. A. Campbell, *DDT and Newer Persistent Insecticides* (Chemical Publishing Company, New York, 1952).

20. There is no completely satisfactory way of appraising the persistence of DDT in nature; residues have different degrees of persistence in different places. The concept of "half-life" implies a systematic degradation that may not occur universally. S. G. Herman, R. L. Garrett, and R. L. Rudd [in *Chemical Fallout*, M. W. Miller and G. G. Berg, Eds. (Thomas, Springfield, Ill., 1969), p. 24] show that the population of western grebes of Clear Lake, California, included individuals having substantially the same range of concentrations of DDD residues over a 10-year period despite the cessation of spraying; mean values dropped by about one-half in that period. Residues are known to remain for many years in soil, especially in organic soils. See the review by C. A. Edwards (21); see also G. M. Woodwell and F. T. Martin (22), J. Dimond (23), and R. G. Nash and E. A. Woolson (24). These observations all suggest that residues persist with mean lives of many years. The assumption of a 10-year half-life for residues within the biosphere as a whole has long appeared reasonable to one of us (G.M.W.) (16); the assumption may be error in that residues tend to be stored or cycled in places where they are not degraded chemically and may persist longer.

21. C. A. Edwards, *Residue Rev.* 13, 83 (1966).

22. G. M. Woodwell and F. T. Martin, *Science* 145, 481 (1964).

23. J. Dimond, *Maine Agr. Exp. Sta. Misc. Rep.* 125 (1969).

24. R. G. Nash and E. A. Woolson, *Science* 157, 924 (1967).

25. U.S. Tariff Commission, *Synthetic Organic Chemicals, U.S. Production and Sales, 1969* (Government Printing Office, Washington, D.C., 1970); *U.S. Dep. Agr. Econ. Rep. No. 158* (April 1969).

26. R. Carson recognized in 1963 the significance of aerial transport of DDT residues in contaminating the oceans. She cited observations in Maine and elsewhere indicating that only half the spray emitted from aircraft lands on the ground, the rest being dispersed in the atmosphere. Recent measurements of DDT attached to particulate matter in the atmosphere confirm the earlier conclusions (R. Carson, statement to the Subcommittee on Reorganization and International Organizations of the Committee on Government Operations, U.S. Senate, May–June 1963, part 1, p. 207); G. M. Woodwell, *Forest Sci.* 7, 194 (1961); R. W. Risebrough, R. J. Huggett, J. J. Griffin, E. D. Goldberg (9); O. B. Cope, *Trans. Amer. Fish. Soc.* 90, 239 (1961).

27. Most measurements have been in agricultural soils where there was a real question about hazards after long use. Published data are, therefore, heavily skewed toward high values, and any tabulation such as that of Table 1 may be misleading. The data of Table 1 are representative of published reports: H. Cole *et al.* (28); E. P. Lichtenstein (29); W. L. Seal *et al.* (30); E. F. Taschenberg *et al.* (31); L. C. Terriere *et al.* (32); W. L. Trautman *et al.* (33); G. W. Ware *et al.* (34); U.S. Department of Agriculture (35). The maximum in this tabulation was in an orchard in Indiana that contained DDT residues totaling 22.0 grams per square meter. Almost all agricultural soils contain detectable DDT residues, most in excess of 0.056 gram per square meter. Averages of several samples from a region are rarely less than 0.224 gram per square meter, according to this tabulation. Nonetheless, Table 1 contains few data from crops that are rarely sprayed, such as many grains and fodder crops. Any average calculated for agricultural soils

must account for the extensive acreages devoted to these crops as well. The average contamination of agricultural land is probably greater than the mean reported for 41 randomly sampled agricultural and forest soils in Wisconsin and eight states west of the Mississippi River (Table 1). In that tabulation, 22 of the soils contained less than 0.0022 gram of DDT per square meter, but others had high enough concentrations to produce an average concentration of 0.148 gram per square meter. It seems very unlikely that the average amount of DDT in agricultural soils, including all agricultural soils, would approach the value of 0.99 gram per square meter reported for soybeans in 1968 (Table 1). A reasonable estimate of the average value for agricultural soils where DDT is used would seem to be in the range of 0.140 to 0.56 gram per square meter. The average would be skewed toward the lower end of the range by the inclusion of crops that are sprayed irregularly. We have assumed for our calculations an average contamination of agricultural soils of 0.168 gram per square meter.

28. H. Cole, D. Barry, D. E. H. Frear, A. Bradford, *Environ. Contam. Toxicol.* **2**, 127 (1967).
29. E. P. Lichtenstein, *J. Econ. Entomol.* **50**, 545 (1957).
30. W. L. Seal, L. H. Dawsey, G. E. Cavin, *Pesticide Monit. J.* **1**, 22 (1967).
31. E. F. Taschenberg, G. L. Mark, F. L. Gambrell, *Agr. Food Chem.* **9**, 207 (1961).
32. L. C. Terriere, U. Kiigemagi, R. W. Zwick, P. H. Westigard, in *Organic Pesticides in the Environment*, R. F. Gould, Ed. (American Chemical Society, Washington, D.C., 1966), p. 263.
33. W. L. Trautman, G. Chesters, H. B. Pionke, *Pesticide Monit. J.* **2**, 93 (1968).
34. G. W. Ware, B. J. Estensen, W. P. Cahill, *ibid.*, p. 129.
35. U.S. Department of Agriculture, Plant Pest Control Division, Agricultural Research Service, *ibid.*, p. 58.
36. Most of the nonagricultural soils for which data on DDT residues are available are also soils from areas that have received repeated applications of DDT. The highest values range from several tenths of a gram per square meter for a marsh along the eastern coast [G. M. Woodwell, C. F. Wurster, Jr., P. A. Isaacson, *Science* **156**, 821 (1967)] and for other organic soils that had been heavily treated, to zero, or at least to values below the limits of detection. Unsprayed forested areas in Pennsylvania (Table .1) contained in 1967 4.5×10^{-4} gram per square meter. Soils in Maine forests that had never been sprayed contained ten times that quantity (Table 1). Organic soils and sediments contain more residues, but usually not more than a few tenths of a pound per acre. Any world average estimated from such data is most tenuous. We estimate that nonagricultural soils contain a minimum of 1.1×10^{-4} gram per square meter. A reasonable estimate of the mean concentration of DDT in the temperate zone of the Northern Hemisphere appears to be 4.5×10^{-4} gram per square

meter as reported for the soils of Pennsylvania forests (Table 1).
37. *Hammond Citation World Atlas* (Hammond, Maplewood, N.J., 1966), p. 192.
38. C. A. Edwards, *Persistent Pesticides in the Environment* (Chemical Publishing Company, New York, 1970).
39. G. S. Nazarov, *Tr. Saratov. Zootekh. Vet. Inst.* **7**, 319 (1958).
40. W. F. Spencer and M. M. Cliath, *Environ. Sci. Technol.* **3**, 670 (1969).
41. "Net primary production" usually refers to the net amount of dry matter produced per unit of land area: G. M. Woodwell and R. H. Whittaker, *Amer. Zool.* **8**, 19 (1968); E. P. Odum, *Fundamentals of Ecology* (Saunders, Philadelphia, ed. 3, 1971); G. M. Woodwell, *Sci. Amer.* **223**, 64 (September 1970).
42. L. H. Long, Ed., *World Almanac* (Newspaper Enterprise Association, Inc., New York, 1967), p. 273; A. M. Piper, *U.S. Geol. Surv. Water Supply Pap. No. 1797* (1965).
43. L. Weaver, C. G. Gunnerson, A. W. Breidenbach, J. Lichtenberg, *Public Health Rep.* **80**, 481 (1965).
44. A. Standen, Ed., *Encyclopedia of Chemical Technology* (Interscience, New York, ed. 2, 1966), vol. 11, p. 691.
45. K. R. Tarrant and J. O. G. Tatton, *Nature* **219**, 725 (1968); G. A. Wheatley and J. A. Hardman, *ibid.* **207**, 486 (1965).
46. J. L. George and D. F. H. Frear, *J. Appl. Ecol.* **3** (Suppl.), 155 (1966).
47. H. Craig, *Tellus* **9**, 1 (1957).
48. R. Revelle and H. E. Suess, *ibid.*, p. 18.
49. W. D. Garrett, *Deep-Sea Res.* **14**, 221 (1967); D. B. Seba and E. F. Corcoran, *Pesticide Monit. J.* **3**, 190 (1969).
50. H. Lipne and C. W. Kearns, *J. Biol. Chem.* **234**, 2129 (1959); L. Goldberg, in *The Earth as a Planet*, G. P. Kuiper, Ed. (Univ. of Chicago Press, Chicago, 1954), p. 434; L. R. Koller, *Ultraviolet Radiation* (Wiley, New York, ed. 2, 1965); A. R. Mosier, W. D. Guenzi, L. L. Miller, *Science* **164**, 1083 (1969); N. Bhandari, D. Lal, Rama, *Tellus* **18**, 391 (1966).
51. W. Wooster, quoted by H. Craig (*47*).
52. W. E. Odum, G. M. Woodwell, C. F. Wurster, *Science* **164**, 576 (1969).
53. G. M. Woodwell, *ibid.* **168**, 429 (1970).
54. H. L. Harrison, O. L. Loucks, J. W. Mitchell, D. F. Parkhurst, C. R. Tracy, D. G. Watts, V. J. Yannacone, Jr., *ibid.* **170**, 503 (1970).
55. M. H. A. Keenleyside, *Can. Fish Cult.* **24**, 17 (1959); *J. Fish. Res. Board Can.* **24**, 807 (1967); F. P. Ide, p. 769; R. B. Anderson and W. H. Euerhart, *Trans. Amer. Fish. Soc.* **95**, 160 (1966); R. B. Anderson and O. Fenderson, *J. Fish. Res. Board Can.* **27**, 1 (1970); R. B. Anderson, *Sebago's Bright Future* (Maine Department of Inland Fisheries and Game, Augusta, 1966); S. DeRoche, unpublished data on fish size in Sebago Lake, 1957–1967, taken by the Maine Department of Inland Fisheries and Game; R. L. Rudd, *Pesticides and the Living Landscape* (Univ. of Wisconsin Press, Madison, 1964).
56. Counts of nesting pairs and young by D. Puleston [Brookhaven Lecture Series No. 104, *Brookhaven Nat. Lab. Publ. 50309* (15 Sep-

tember 1971)] and his colleagues over more than 20 years have documented the catastrophic decline of the osprey in the early 1960's, and the slight recovery of reproductive success in recent years. Whether the small residual population will recover remains doubtful, however.

57. D. A. Ratcliffe, *J. Appl. Ecol.* **7**, 67 (1970).
58. W. J. L. Sladen, C. M. Menzie, W. L. Reichel, *Nature* **210**, 670 (1966).
59. C. W. Stanley, J. E. Barney, II, M. R. Helton, A. R. Yobs, *Environ. Sci. Technol.* **5**, 430 (1971).
60. R. H. Whittaker, *Communities and Ecosystems* (Collier-Macmillan, New York. 1970).
61. H. J. M. Bowen, *Trace Elements in Biochemistry* (Academic Press, New York, 1966).
62. Research was carried out under the auspices of the U.S. Atomic Energy Commission.

SITES OF INHIBITION OF PHOTOSYNTHETIC ELECTRON TRANSPORT BY 1,1,1-TRICHLORO-2,2-BIS-(P-CHLOROPHENYL)ETHANE (DDT)

L.J. ROGERS, W.J. OWEN AND M.E. DELANEY

The reaction of barley to DDT (fig. 1) is controlled by a single major gene with susceptibility dominant to resistance (HAYES 1959, WIEBE & HAYES 1960). The susceptible varieties develop a severe chlorosis some 4-10 days after treatment with DDT whereas similarly treated resistant varieties are not affected.

Previous studies by LAWLER & ROGERS (1967, 1968, 1969) have shown that DDT affects the photosynthesis of susceptible barley, as evidenced by a decrease in rate of O_2 evolution of DDT-treated leaves and diminished Hill activity of the chloroplasts isolated from treated plants. In chloroplasts isolated by a procedure adapted from KLEESE (1966) DCPIP photoreduction and phenazine methosulphate-catalysed cyclic photophosphorylation were inhibited, whereas there appeared to be little effect on ferricyanide photoreduction or concomitant non-cyclic photophosphorylation.

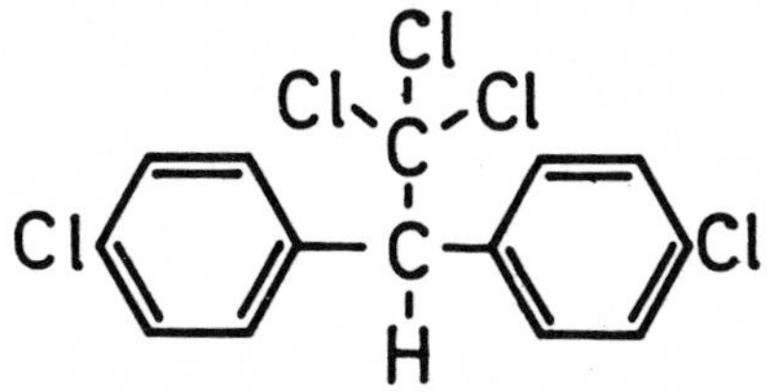

Fig. 1: The structure of DDT. Analogues of DDT with (a) $-OCH_3$, Br, Cl or F in the para-positions of the phenyl-ring; (b) the C2 ethane hydrogen; and (c) the trichloro- or tribromo-ethane grouping are also toxic to some varieties of barley.

Recently (OWEN et al. 1970) we have studied the effect of DDT on these biochemical functions in some 30 varieties of barley from widely differing parts of the world [e.g. Fimbul (Finland), Sinop 248 (Turkey), Lyallapur (India), Tang May (China), Mosdoksky H.V. (Russia), Stade Paladi (Barbados), Atlas 46 (U.S.A.)]. In all cases the effect of DDT on photosynthesis correlated with the susceptibility or resistance of the barley to DDT in greenhouse trials. In these studies chloroplasts from susceptible barley showed a parallel effect on all the investigated aspects of photosynthesis, including DCPIP and ferricyanide photoreductions, and cyclic and non-cyclic photophosphorylations (Table I). The distinction between cyclic and noncyclic photophosphorylations reported in earlier work is peculiar to only one or two susceptible varieties following preparation of chloroplasts in certain media before assay of photosynthetic activities.

In subsequent studies reported here we have investigated the effect of DDT on photosynthetic electron transport. Susceptible and resistant types of a single variety of barley (Zephyr) were used throughout; the

147

two types of barley appear to be identical but for the distinction in their reaction to DDT. The plants were treated at the two leaf stage with a DDT emulsion (0.5 g DDT in 20 ml acetone, then 0.1% Tween 60 in H_2O to 200 ml) delivered as a spray. In all experiments control (untreated) plants were sprayed with a similar mixture not containing DDT. Chloroplasts were isolated from plants some 48 hr after DDT-treatment, well before chlorosis began to develop. All studies made on DDT-susceptible barley were accompanied by experiments using the resistant type. In these latter studies there was no observed inhibition resulting from DDT treatment in any of the investigated aspects of photosynthetic activity; this data is therefore not presented.

Electron flow pathways in isolated chloroplasts are associated with one or with both photosystems and may be of a cyclic or a non-cyclic nature. The light induced reduction of $NADP^+$ by water requires the participation of both photosystems 1 and 2, operating in series and joined by an electron transport chain involving plastoquinone, cytochrome b_{559}, cytochrome f, and plastocyanin (Fig. 2). For each four electrons transported one molecule of O_2 is evolved and two molecules of $NADP^+$ reduced.

In an attempt to localise the site of action of DDT, experiments utilising artificial electron donors and acceptors were carried out. These data, now reported briefly, indicate that DDT inhibits photosynthetic electron transport at two sites, a site before photosystem 2 and

Table I. Effect of DDT on photoinduced electron transport and photosynthetic phosphorylations in barley.

Variety	Hill activity				Photophosphorylations			
	DCPIP		Ferri-cyanide		Cyclic		Non-cyclic	
		DDT		DDT		DDT		DDT
Bohmerweld	110	110	145	145	1155	1230	495	465
Zephyr DDT-resistant	95	105	130	130	1230	1380	255	350
Zephyr DDT-susceptible	70	20	145	40	1420	470	290	25
Klintso	85	25	130	65	1430	280	260	55

Plants were sprayed and chloroplasts isolated two days later before chlorosis began to develop in the susceptible type. Results are given for DDT-treated plants and untreated controls.
Experimental details are given briefly elsewhere (OWEN et al. 1970). Rates are given as µmoles electron acceptor reduced/mg-Chl/hr or µmoles ATP formed/mg-Chl/hr as appropriate. Because of limits of space data for only two susceptible (Klintso and Zephyr) and two resistant (Bohmerweld and Zephyr) varieties are presented.

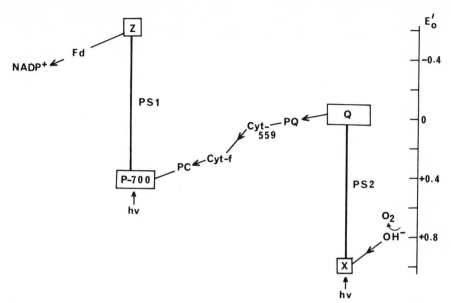

Fig. 2: Present concepts of photoinduced electron flow in photosynthesis in higher plants. Direction of flow is indicated by the arrows. The review by BOARDMAN (1968) should be consulted for detailed explanation. (PQ, plastoquinone; PC, plastocyanin; Fd, ferredoxin; PS1, photosystem 1; PS2, photosystem 2).

a second site in the intermediate electron transport chain linking the two photosystems.

In chloroplasts isolated from DDT treated susceptible barley both DCPIP and ferricyanide photoreduction were inhibited (see e.g. Table I) indicating that a site of inhibition by DDT was associated with electron transport prior to photosystem 2 or with the early electron carriers of the intermediate electron transport chain.

In subsequent studies a site of inhibition of electron transport by DDT before photosystem 2 was identified from spectrophotometric studies on tris-washed chloroplasts in which electron donation from water, and O_2 evolution, have been eliminated (YAMASHITA & BUTLER 1968). In this system artificial electron donors such as diphenylcarbazide (DPC) were used to donate electrons beyond the site of tris inhibition (YAMASHITA & BUTLER 1969, VERNON & SHAW 1969) and electron flow through photosystem 2 was measured by use of DCPIP or ferricyanide as electron acceptor from the intermediate electron-transport chain (Fig. 3).

The properties of chloroplasts from DDT-treated susceptible barley, where electron transport from water, as evidenced by photoreduction of DCPIP, was inhibited some 50%, were compared with those from untreated controls. In both cases washing with tris almost completely inhibited electron donation from water, although electron transport could be restored to the control (untreated barley) value by diphenylcarbazide (Fig. 4). Thus with electron transport from diphenylcarbazide no inhi-

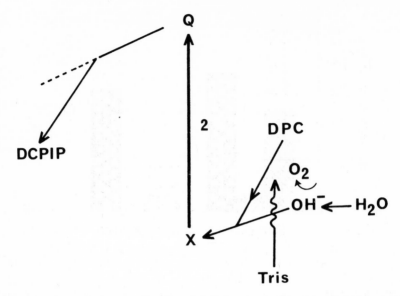

Fig. 3: Photoinduced electron flow from 1,5-diphenylcarbazide through photosystem 2 to DCPIP. Electron donation from water, and associated O_2 evolution, have been eliminated by treatment with tris.

bition of electron transport to DCPIP was observed in chloroplasts from DDT-treated barley. This indicates that a site blocked by DDT in the electron transport chain before photosystem 2 is by-passed by electron donation from diphenylcarbazide. In contrast with work on spinach chloroplasts no electron donation beyond the tris block from benzidine or semicarbazide (YAMASHITA & BUTLER 1969) was observed.

The observation (LAWLER & ROGERS 1968, OWEN et al. 1970) that DDT affects cyclic photophosphorylation, a photosynthetic activity that does not involve photosystem 2, indicates there is probably also inhibition at some other site. From recent studies this second site of inhibition by DDT appears to be located in the intermediate electron transport chain linking the two photosystems.

This site was identified from studies of electron donation to photosystem 1 by DCPIPH$_2$-ascorbate, TMPD-ascorbate or DAD-ascorbate couples. Electron donation was measured in the oxygen electrode by O_2 consumption using diquat or paraquat as electron acceptor from photosystem 1 (IZAWA et al. 1966, BÖHME & TREBST 1969). Electron donation from water through photosystem 2 was inhibited by DCMU (Fig. 5).

Chloroplasts were isolated from DDT-treated susceptible barley some three days after spraying with DDT, when Hill activity, as measured by DCPIP or ferricyanide photoreduction, was inhibited some 50%. This inhibition is a measure of the affect of DDT at the site prior to photosystem 2.

In such chloroplasts electron donation from DCPIPH$_2$ was the same as in chloroplasts isolated from untreated plants; however, when TMPD or

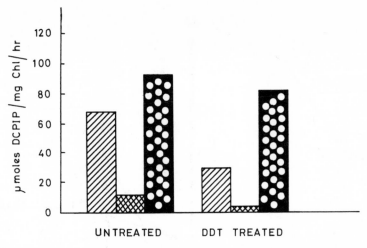

Fig. 4: Inhibition by DDT of photoinduced electron flow through photosystem 2 from
water, and restoration in tris-washed chloroplasts by electron donation from DPC, as
shown in Fig. 3.
Chloroplasts were isolated in 0.05 M tris (pH 7.8) containing 0.4 M sucrose and
0.01 M NaCl. Where indicated chloroplasts were washed in 0.3 M tris (pH 8.0) for 15
min. at approx. 4°C. The incubation mixture for assay of electron transport contained
68 μmoles KH_2PO_4, 0.25 μmoles DPC, 0.16 μmoles DCPIP in a total volume of 3.0 ml. of
0.16 M sucrose. Chloroplasts equivalent to 40 μg chlorophyll were used in assays.
Temp. 20°C. 5000 ft-c. illumination.
At the conclusion of illumination periods 0.1 ml 1.7×10^{-4} M DCMU was added to each
cuvette to check that electron transport through photosystem 2 accounted for the observed photoreduction of DCPIP. Rates are expressed as μmoles DCPIP reduced/mg chlo-
rophyll/hr. Results are shown for chloroplasts isolated from DDT-treated and untreated susceptible barley. Cross-hatched areas, rate of electron transport from
water to DCPIP ; hatched areas, rate of electron transport from water to DCPIP after
tris-washing; circled areas, rate of electron transport from DPC to DCPIP in tris-
washed chloroplasts.

DAD were used as electron donors electron transport was significantly
lower, the inhibition being comparable to that observed for the site
prior to photosystem 2. Under comparable conditions electron donation
from DAD was less than that from TMPD (Fig. 6). These data indicate
that there is a site of inhibition by DDT in the intermediate electron
transport chain before the site of electron donation from $DCPIPH_2$ but
after the sites of electron donation from TMPD or DAD. We have observed
(Table I), that DDT will inhibit phenazine methosulphate catalysed cyclic photophosphorylation. Unless there is a further, as yet unidentified, site of inhibition by DDT the results appear to indicate that in
cyclic electron flow electron return to the intermediate electron transport chain is at a level before (more negative E_o^1 than) the site of
electron donation from $DCPIPH_2$.
    When chloroplasts were isolated from susceptible barley and then
treated with 400 μg DDT/mg chlorophyll for 60 min no inhibition of

151

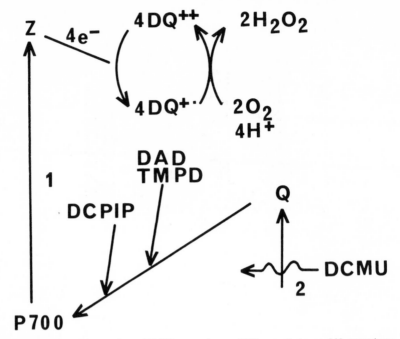

Fig. 5: Electron transport from DCPIPH₂-ascorbate, TMPD ascorbate or DAD-ascorbate couples through photosystem 1 to diquat. Electron donation from water through photosystem 2 is inhibited by DCMU. The stoichiometry shown is that for transfer of four electrons.

electron donation from DCPIPH$_2$, TMPD or DAD to diquat was shown even under conditions where Hill activity as measured by DCPIP photoreduction was 45% inhibited. In previous studies of other photosynthetic activities (LAWLER & ROGERS 1968, OWEN et al. 1970, DELANEY et al. 1971) results have been similar for either type of DDT treatment. The reason for the anomalous results in the case of electron donors to the intermediate electron transport chain is not clear though it may be that following attachment of DDT to the 'target site' in a lamellar membrane a precise spacial orientation of the DDT molecule is necessary before interference with electron transport results. When chloroplasts are isolated from DDT-sprayed plants sufficient time has elapsed for this to happen whereas in isolated chloroplasts treated with DDT this does not readily occur. The orientation of DDT necessary to inhibit electron flow at the site before photosystem 2 may not be so critical since inhibition is readily obtained with chloroplasts treated *in vitro* with DDT.

Studies of electron donation by Mn^{2+} or ascorbate to the electron-transport chain before photosystem 2 (BEN-HAYYIM & AVRON 1970) were also made by measuring O$_2$ consumption in the oxygen electrode with diquat as electron acceptor from photosystem 1 (IZAWA et al. 1966, BÖHME

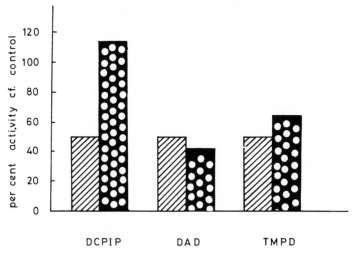

Fig. 6: Photoinduced electron flow through photosystem 1 from DCPIPH$_2$, TMPD and DAD, as shown in Fig. 5. Chloroplasts were isolated from DDT-treated and untreated susceptible barley in 0.05 M KHCO$_3$ (pH 7.5) containing 0.5 M sucrose. The incubation mixture for assay of electron transport contained 200 μmoles TES buffer (pH 7.0); 8 μmoles MgCl$_2$; 0.068 μmoles DCMU; 12 μmoles Na-ascorbate; 0.4 μmoles diquat, and 0.4 μmoles DCPIP or 2.0 μmoles TMPD or DAD in a total volume of 4.0 ml. Chloroplasts equivalent to 50-80 μg chlorophyll were used in assays. Temp. 20°C. 5000 ft-c.illumination. Rates are given as percent activity compared to chloroplasts from untreated barley. Results are shown for electron donation from DCPIPH$_2$, TMPD and DAD. Hatched area, electron transport from water to DCPIP through photosystem 2; circled area, electron transport from the added donor to diquat through photosystem 1. The rates of electron transport for untreated controls were: DCPIP photoreduction, 130 μmoles/ mg chlorophyll/hr.; DCPIPH$_2$; TMPD; DAD; 150; 520; 800 μmoles O$_2$/mg chlorophyll/hr. respectively.

& TREBST 1969) (Fig. 7). Mn^{2+} or ascorbate can compete successfully with water as electron donors to photosystem 2. Since, in the presence of Mn^{2+} or ascorbate, photosynthetic O$_2$ evolution is thus abolished a doubling in the rate of O$_2$ uptake due to electron acceptance from photosystem 1 by diquat is observed (Fig. 8). In chloroplasts prepared by the methods used insignificant levels of catalase appeared to be present. However, azide was included in the incubation medium as an additional precaution to eliminate any loss of observed oxygen uptake due to peroxide decomposition and resultant release of molecular oxygen.

In these studies no change was observed in relative rates of electron donation from the artificial donors compared with donation from water in chloroplasts isolated from DDT-treated or untreated susceptible barley (Table II). This may suggest that the sites of electron donation from Mn^{2+} or ascorbate are before the target site for DDT in the electron transport chain before photosystem 2, since in chloroplasts from treated barley some alleviation of the observed inhibition of electron transfer from water might be expected if the DDT-sensitive site were by-passed. Therefore, it is concluded that the inhibition ob-

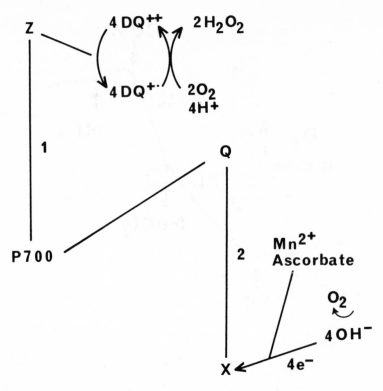

Fig. 7: Photoinduced electron flow from Mn^{2+} or ascorbate through photosystem 2 and photosystem 1 to diquat. The stoichiometry shown is that for transfer of four electrons.

served in electron donation from Mn^{2+} or ascorbate is a function of inhibitions by DDT of both the site before photosystem 2 and the site in the intermediate electron transport chain (Fig. 9).

Current investigations on aspects of electron transport are concerned with studies on the protection of the DDT-sensitive site by prior treatment with structural analogues of DDT, such as 1,1-dichloro-2,2-bis(p-chlorophenyl)ethylene (DDE), which are not toxic to DDT-susceptible barley. It will be interesting to see whether both sites behave similarly in this respect. Other studies indicate that the chloroplast thylakoid proteins isolated from susceptible and resistant types of Zephyr barley differ slightly in amino acid composition; it may be that some amino acid sequence(s) unique to the susceptible barley constitute an integral part of the 'target-site' for DDT in the thylakoid membranes.

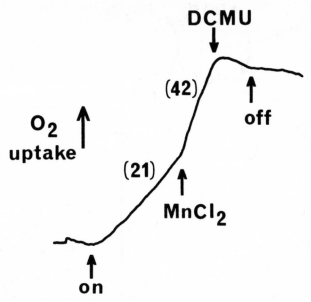

Fig. 8: Effect of Mn^{2+} on O_2 uptake in the presence of diquat. Typically the reaction mixture contained 45 µmoles Tricine (pH 8.0), 2.0 µmoles $MgCl_2$, 0.1 µmoles diquat, 3.0 µmoles NaN_3, and chloroplasts containing 40-60 µg chlorophyll, in a total volume of 3.0 ml. Temp. 20°; 5000 ft-c. illumination. Numbers in parentheses represent rates of O_2 uptake expressed in µmoles/mg chlorophyll/hr. The example given is that for chloroplasts from untreated plants (see Table II).
For ascorbate the reaction mixture was similar except that ascorbate (4.0 µmoles) was added instead of $MnCl_2$.

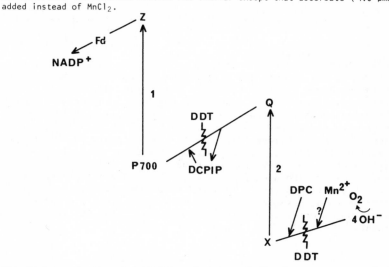

Fig. 9: Apparent sites of inhibition of photosynthetic electron transport by DDT.

155

Table II. Effect of Mn^{2+} and ascorbate O_2 uptake in the presence of diquat for DDT-treated and untreated susceptible barley.

Source of chloroplasts	Electron donor	Untreated	DDT-treated	% Inhibition of electron transport
A	H_2O	21	8	62
	Mn^{2+}	42	12	71
A	H_2O	10	4	60
	Ascorbate	38	12	69
B	H_2O	24	11	54
	Mn^{2+}	46	23	50
B	H_2O	30	18	40
	Ascorbate	68	40	40

Experimental details are indicated in Fig. 8.
A - experiments with chloroplasts from DDT-treated and untreated plants;
B - experiments with chloroplasts from untreated plants incubated with
    200 μgm DDT/mg chlorophyll for 60 min. prior to assay.
Rates are μmoles O_2/mg chlorophyll/hr.

Acknowledgements

The authors thank Dr. J.D. HAYES (Welsh Plant Breeding Station, Aberystwyth) for supplies of seeds and valuable discussion. L.J.R. gratefully acknowledges receipt of a grant from the Agricultural Research Council supporting the work.

Abbreviations

DDT, 1,1,1-trichloro-2,2-bis-(p-chlorophenyl)ethane; DCPIP, 2,6-dichlorophenol-indophenol; TMPD, N,N,N',N'-tetramethyl-p-phenylenediamine; DAD, 2,3,5,6-tetramethyl-p-phenylenediamine (also known as diaminodurene); DPC, 1,5-diphenylcarbazide; diquat, 1,1'-ethylene-2,2'dipyridylium dibromide; paraquat, 1,1'dimethyl-4,4'-dipyridylium dichloride; DCMU, 3-(3,4-dichlorophenyl)-1,1-dimethylurea.

Bibliography

BEN-HAYYIM, G. & B. AVRON - 1970 - *Biochim. Biophys. Acta*, 205, 86.
BOARDMAN, N.K. - 1968 - *Adv. Enzymol.*, 30, 1.
BÖHME, H. & A. TREBST - 1969 - *Biochim. Biophys. Acta*, 180, 137.
DELANEY, M.E., W.J. OWEN & L.J. ROGERS - 1971 - *Biochem. J.*, 124, 24P.

HAYES, J.D. - 1959 - *Nature*, 183, 551.
IZAWA, S., T.N. CONNOLLY, G.D. WINGET & N.E. GOOD - 1966 - Brookhaven
  Symp. Biol., 19, 169.
KLEESE, R.A. - 1966 - *Crop Sci.*, 6, 524.
LAWLER, P.D. & L.J. ROGERS - 1967 - *Nature*, 215, 1515.
LAWLER, P.D. & L.J. ROGERS - 1968 - *Biochem. J.*, 110, 381.
LAWLER, P.D. & L.J. ROGERS - 1969 - Progress in Photosynthesis Research,
  Vol. 3, Ed. by H. METZNER, International Union of Biological Sciences,
  1969, p. 1761.
OWEN, W.J., L.J. ROGERS & J.D. HAYES - 1970 - *Biochem. J.*, 121, 6P.
VERNON, L.P. & E.R. SHAW - 1969 - *Pl. Physiol.*, 44, 1645.
WIEBE, G.A. & J.D. HAYES - 1960 - *Agron. J.*, 52, 685.
YAMASHITA, T. & W. BUTLER - 1968 - *Pl. Physiol.*, 43, 1918.
YAMASHITA, T. & W. BUTLER - 1969 - *Pl. Physiol.*, 44, 435.

DDT Residues in Marine Phytoplankton: Increase from 1955 to 1969

JAMES L. COX

Annual use of DDT in the United States has declined in the past decade (*1*), yet there is recent evidence of abnormally high DDT residues in marine fish from U.S coastal waters, and such contamination in these areas may exceed that in freshwater habitats (*2*). This could indicate either (i) that environmental DDT residues are increasing or (ii) that these recent analyses simply reflect current DDT input and that DDT concentrations have in fact been even higher in the past. Although DDT residues in estuarine shellfish (*3*) have shown no consistent upward or downward trends, the time has been too short and the estuarine system too responsive to weather conditions and local sources of pesticides to provide any measure of the trends in the coastal environment. Declining reproductive success in species of marine pelagic birds, attributable to DDT residues (*4*), does suggest that residues of DDT are increasing in the coastal pelagic food chains of which these birds are high-order consumers.

A decision between the alternatives could be made if historical collections of marine organisms were available. At the Hopkins Marine Station, sam-ples (composed primarily of phytoplankton) collected with a fine-mesh net from Monterey Bay, California, have been collected from 1955 to 1969 (*5*). Phytoplankton samples are particularly suited for analysis because they represent the first link in pelagic food chains. Trends in their concentrations of DDT residues are relevant to all higher-order consumers on the food chain. Also, DDT uptake by phytoplankton is rapid and essentially irreversible (*6*); thus, it can be assumed that the content of DDT residues of phytoplankton reflect prevailing amounts of environmental DDT. To examine the change in content of DDT residues over the collection period, 23 samples from the collection were analyzed. All the samples had been preserved in a 3 percent solution of formalin in seawater. The estimated concentrations of DDT residues (*7*) for the samples were based on their carbon content as determined by wet combustion (*8*) of replicate portions. Formalin induces error in carbon determinations of marine planktonic material (*9*), but the errors in this instance were small (< 10 percent). Treatment of freshly collected material from the same station with formalin had no apparent effect on estimates of the DDT content when compared to that of frozen controls.

Samples were filtered onto combusted GFC glass-fiber filters (Whatman) after filtration through 0.33-mm netting to remove larger zooplankton. The sample and filter pad were ground together in three successive rinses of high-purity *n*-hexane. The pooled rinses were concentrated and chromatographed on silica-gel microcolumns (*10*). Eluates from the columns were concentrated at 37°C under a stream of nitrogen and analyzed by gas-liquid chromatography (GLC). All glassware used in the procedure was combusted at 350°C overnight prior to use; this treatment reduced background contamination nearly to zero for the GLC analyses. Recovery from a variety of samples of known content of DDT exceeded 95 percent in all cases.

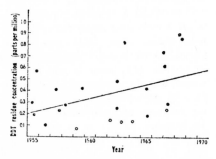

Fig. 1. Concentrations of DDT residues (*7*) in samples of phytoplankton collected by towed nets from Monterey Bay, California, 1955 to 1969. Concentrations are expressed as weight of estimated DDT residues per unit wet weight of phytoplankton as converted from measurements of oxidizable organic carbon content of the samples (*16*). Solid circles indicate smallest samples (<0.2 mg of carbon); half-solid circles indicate samples with 0.2 to 0.4 mg of carbon; open circles indicate samples with greater than 0.4 mg of carbon.

The extracts were injected into a Beckman GC-4 gas chromatograph equipped with two columns and two electron-capture detectors (*11*). Each sample was chromatographed on at least two columns of different composition (*12*). Peaks were identified by standard injection retention times, fortified injections, and disappearance of presumptive DDT and DDD peaks caused by treatment of the extracts with KOH in alcohol.

There were higher concentrations of DDT residues in more recent samples (Fig. 1). Inasmuch as sample storage may have affected this apparent temporal trend, experiments were performed to test the effect of decomposition on the relative proportions of the three DDT constituents found in the samples. Ring-labeled ^{14}C-DDT was added to sealed ampules that contained portions of a phytoplankton sample preserved in formalin. The amounts of labeled compound added were comparable to those amounts found in the 23 analyzed samples. Because the samples had been stored in the dark, it was assumed that any possible breakdown would be thermochemical, not photochemical. Elevated temperatures for short periods of time were used to recreate longer periods at room temperature. The contents were heated at 30°C, 60°C, and 75°C for 6 days and then removed from the ampules, extracted, and analyzed by thin-layer chromatography (*13*). Narrow (0.5-cm) zones were scraped from the chromatoplates into scintillation vials for measurements of ^{14}C activity. The degree to which DDT was broken down to polar compounds affected the relative proportions of the remaining nonpolar constituents: *p,p'*-DDT, *p,p'*-DDD, and *p,p'*-DDE. No breakdown of the ^{14}C-DDT occurred in the samples heated at 30°C. In the samples heated at 60°C,

28 percent of the ^{14}C-DDT broke down to polar compounds, but about half of the remaining nonpolar material was p,p'-DDT. In the sample heated at 75°C, 38 percent of the ^{14}C-DDT broke down to the polar compounds, but p,p'-DDT comprised only 15 percent of the nonpolar material, whereas p,p'-DDD and p,p'-DDE comprised 83 percent. On the basis of these experiments, a change in the relative proportions of the DDT residues in a sample would be expected if any net decomposition to nonpolar products had occurred during storage. In fact, the relative proportions of the DDT constituents found in the samples were quite constant (14); percent regression analysis showed the slope of each percentage versus time function to be not significantly different from zero. Therefore, the trend indicated in Fig. 1 represents an actual increase in the DDT residues in the phytoplankton rather than a loss of analyzable residues during sample storage.

Part of the variability in the values in Fig. 1 is attributable to sample size. Due to the nature of the collection technique, the sample sizes were directly related to the density of the phytoplankton in the water at the time of collection. The estimates of carbon content in the 23 samples were based on standard portions of the samples, which in turn contained the entire contents of a vertical ¼-meter net tow from a depth of 15 meters to the surface. The carbon values are thus indices of standing crop density (Fig. 2). The assumption behind the theoretical curve (Fig. 2) is that a fixed amount of pesticide residue becomes incorporated in the algal material present in a given volume of water, regardless of the density of the standing crop. However, density of the standing crop

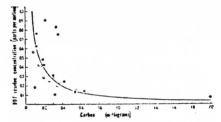

Fig. 2. The effect of size of relative standing crop (milligrams of carbon) on the estimated concentration of DDT residue (7). The theoretical curve was computed according to the relationship $C \times D = k$, where C = carbon content, D = DDT residue concentration and k = weight of the mean amount of DDT residues in the samples. Solid circles indicate samples taken from the later half of the sampling period; open circles indicate samples from the earlier half. Values on the vertical axis were derived as in Fig. 1.

affects the final concentration of acquired residues according to the relationship in Fig. 2. This suggests that the partition coefficient of DDT residues for phytoplankton and similar material (6, 15) diminishes as the density of the phytoplankton increases. A comparison of the points falling above or below the theoretical curve in Fig. 2 shows that the preponderance of later points have higher concentration values, despite the effect of the size of standing crop. The same conclusion may be reached by examining the points in Fig. 1 by size classes.

The residues of DDT may be increasing in the primary stages of coastal pelagic food chains. If the processes of decomposition and dispersal of these residues in succeeding steps are not sufficiently rapid to counteract this apparent increase, a delay may be expected before the decline of domestic usage of DDT begins to be reflected in the components of these food chains.

References and Notes

1. T. Eichers and R. Jenkins, *U.S. Dep. Agr. Agr. Econ. Rep. No. 158* (1969).
2. R. W. Risebrough, *Chemical Fallout, First Rochester Conference on Toxicity* (Thomas, Springeld, Ill., 1969), p. 8; also analyses contracted by the California Department of Public Health showed some samples of canned jack mackerel *Trachurus symmetricus*, Ayres, from California waters had in excess of 5.0 parts of DDT residues per million.
3. P. A. Butler, *Bioscience* 19, 889 (1969); *U.S. Bur. Comm. Fish. Res. Contract Rep. 14-17-0002-265* (1970).
4. R. W. Risebrough, D. B. Menzel, D. J. Martin, Jr., H. S. Olcott, *Nature* 216, 589 (1967); C. F. Wurster, Jr., and D. B. Wingate, *Science* 159, 979 (1968); D. B. Peakall, *Sci. Amer.* 222, 72 (1970); D. B. Peakall, *Science* 168, 592 (1970).
5. The samples discussed here are part of a larger series which has been described in detail by R. L. Bolin and D. P. Abbott, *Calif. Coop. Ocean. Fish. Invest. Rep.* 9, 23 (1963) and D. P. Abbott and R. Albee, *ibid.* 11, 55 (1967).
6. A. Sodergren, *Oikos* 19, 126 (1967); J. Cox, *Bull. Environ. Contam. Toxicol.* 5, 218 (1970).
7. The DDT residues include the constituents of technical DDT and the nonpolar metabolites derived therefrom. In this study only p,p'-DDT [1,1,1-trichloro-2,2-bis(p-chlorophenyl)ethane], p,p'-DDD [1,1-dichloro-2,2-bis(p-chlorophenyl)ethane], and p,p-DDE [1,1-dichloro-2,2-bis(p-chlorophenyl)ethylene] were detected in measurable amounts. The term DDT residues as it is used in the text refers to all of these three compounds.
8. J. D. H. Strickland and T. R. Parsons, *Bull. Fish. Res. Board Can.* 167, 207 (1968).
9. T. L. Hopkins, *J. Cons. Cons. Perma. Int. Explor. Mer* 31, 300 (1968).
10. A. M. Kadoum, *Bull. Environ. Contam. Toxicol.* 3, 65 (1968); *ibid.*, p. 354.
11. All GLC parameters were those suggested in *Pesticide Analytical Manual* (U.S. Dept. of Health, Education, and Welfare, Food and Drug Administration, revised, 1968). vol. 2.
12. Coatings used on the columns were 5 percent DC-200, 5 percent QF-1, 5 percent mixed bed of DC-200 and QF-1, and 3 percent SE-30 with 6 percent QF-1 in a mixed bed. All coatings were made on DMCS Chromosorb W.
13. Silica gel G was used as the adsorbent. Chromatoplates were developed in *n*-heptane, compounds were identified by cochromatography with pure standards.
14. Values expressed as percent followed by standard error in percent: p,p'-DDT, 57.1 ± 12.9; p,p'-DDE, 18.0 ± 6.7; and p,p'-DDD, 24.9 ± 13.7.
15. C. F. Wurster, Jr., *Science* 159, 1474 (1968).
16. Wet weight concentrations were computed by converting carbon content of the samples to estimated wet weights by multiplying by 100. This factor was derived from values given by E. Harris and G. A. Riley [*Bull. Bingham Oceanogr. Collect. Yale Univ.* 15, 315 (1956)] and H. Curl, Jr. [*J. Mar. Res.* 20, 181 (1962)].
17. Supported by NSF grant GB 8408, a grant from the State of California Marine Research Committee, and an NSF predoctoral fellowship. I thank David Bracher for technical assistance with the carbon and DDT analyses.

KEY-WORD TITLE INDEX

AUTHOR INDEX